DIGITAL PLACES

The last twenty-five years have seen major changes in the nature and scope of geographical information. This has happened in one way in society at large, where computers, satellites and global positioning systems have made geographical information more extensive, more detailed and more available. It has happened in another way within the university, where rapidly evolving geographic information systems have been touted as tools useful in a wide range of disciplines, tools that will resolve problems as different as the nature of global climate change and the routing of mail.

In both settings the move from manual to computer-based systems is viewed as having a natural trajectory, from less powerful to more powerful technologies. These systems are held to be increasingly able to model and represent all that is important in geographical knowledge and behavior. They are seen as fitting into and supporting traditional scientific and social practices and institutions.

Digital Places: Living with Geographic Information Technologies shows that on each score the systems have been misunderstood and their impacts underestimated. By offering an understanding of geographic information systems within the social, economic, legal, political and ethical contexts within which they exist, the author shows that there are substantial limits to their ability to represent the very objects and relationships, people and places, that many believe to be most important.

Focusing on the ramifications of GIS usage, *Digital Places* shows that they are associated with far-reaching changes in the institutions in which they exist, and in the lives of those they touch. In the end they call for a complete rethinking of basic ideas, like privacy and intellectual property and the nature of scientific practice, that have underpinned public life for the last hundred years.

Michael R. Curry is an Associate Professor in Geography at the University of California, Los Angeles.

D1307191

DIGITAL PLACES

Living with geographic information
technologies

Michael R. Curry

London and New York

First published 1998
by Routledge
11 New Fetter Lane, London EC4P 4EE

Simultaneously published in the USA and Canada
by Routledge
29 West 35th Street, New York, NY 10001

Typeset in Perpetua by Routledge
Printed and bound in Great Britain by TJ International Ltd,
Padstow, Cornwall

British Library Cataloguing in Publication Data
A catalogue record for this book is available from the British Library

Library of Congress Cataloguing in Publication Data
Digital Places: living with geographic information / Michael R. Curry
Includes bibliographical references and index.
1. Geographic information systems. I. Title.
G70.212.C87   1998
98–14913 CIP
910′.285–dc21

ISBN 0–415–13014–X (hbk)
ISBN 0–415–13015–8 (pbk)

FOR ANNEMARIE, ETHAN, AND RACHEL

CONTENTS

CONTENTS

ILLUSTRATIONS

Tables

Figures

PREFACE

One of the striking things about geographic information systems is the difficulty that one has when one attempts to talk about them, or at least when one attempts to talk about them in ways that are not strictly technical. There are now a number of people, both within geography and outside of it, who have begun to discuss the social and cultural and ethical issues that the systems raise, and every one of them, at least every one with whom I've spoken, has had experiences similar to the following.

A few years ago I delivered a talk in a geography department, one known as the home of pioneers in geographic information systems. As I finished the presentation a member of the audience, a long-time user of the systems, stood up and started a long questioning monologue, in which he claimed to find fault with almost all of what I had said. He peppered his comments with examples, all aimed at refuting what I had just said, on "Geographic information systems, revolution, and the ethos of science."

As I suggested, I have shared this experience with others. But as he continued, a second thing began to happen, and others, too, have had this experience. It began to strike me that his original question had gradually evolved, turning into the recitation of a litany of problems that the user of a geographic information system faces. Virtually every one of his examples was in fact evidence in support of what I had said.

Why did this person—and as I say, this was not an isolated event—think that we were disagreeing, when it seemed to me that we were very much in agreement? On reflection, it seems to me that what was missing in his reflections, and what is often missing in the reflections provided by users of technologies, is an understanding of the ways in which our work fits into a broader picture. In large measure, what is missing is an appropriate vocabulary. As a result, we tend to view the problems that arise when we use the technology as idiosyncratic and personal. In fact, though, few such problems really are entirely personal or entirely idiosyncratic. By and large, the difficulties and decisions that I face in using a technology have very likely been faced before by others, and will very likely be faced, in the future, by even more.

In looking at geographic information systems, then, the task is not so much to ferret out hidden difficulties and problems. Rather, it is to place the everyday experiences of the users of that technology in a broader context, and thereby to make them visible in a different way. It is to show the ways in which what appears individual and particular is actually general, just as it is to show that what appears general is actually particular. Indeed, it is only by engaging in these tasks that one can begin to understand what options there are, that one can begin to see ways out of what look like impasses.

This work has been long in preparation. In a very real sense I began it in 1977, when I first confronted geographic information systems as a student worker at Minnesota Land Management Information Systems at the University of Minnesota, and then at the Land Management Information Center in the Department of State Planning. A new graduate student, with an interest in culture and society, I immediately saw something of interest. Later, in a couple of years working full-time in a private business that produced computer maps for industry, I saw another side of the GIS business, the side more closely related to geodemographics, and to issues of culture theory, but also of privacy and intellectual property. And as a faculty member, teaching cartography and geographical methodology, I have seen a third side, one related to the impacts of technology on the academy.

But if this work draws very much from my own experiences, it is not simply a personal view. It is not, in the end, a rumination on my own experiences. Rather, it uses those experiences as a springboard for a much broader analysis of geographic information systems. When I say that this is a broader analysis, I actually mean that in three rather different ways. First, it is broader insofar as it involves the attempt to see the systems as a particular instance of technological systems more generally. In a sense, this volume can be seen as addressing the question: How ought we to look at technological systems generally, and what do we find when we narrow our focus to the particular case of geographic information systems? Indeed, here the reader ought to see this work as continuous with my last book, on the nature of the written work in geography.

There is a second way, though, in which this work can be seen as a broader analysis. And that is because I have adopted a broad definition of "geographic information systems." While some would prefer to limit the definition, and for example to make a distinction between the systems and automated cartography, I would argue that one needs to take this broader view.

I take geographic information systems to be technological means for the collection, storage, analysis, and representation of geo-coded data. This definition, to my mind, clearly includes computer-assisted cartography. At the same time, it equally requires the inclusion of technologies like remote sensing and global positioning systems. (And because it does not require that the systems be capable of *systematic* collection etc., it includes—at least in principle—certain means that most would certainly exclude, such as travel guides and community bulletin boards, which will enter into my analysis only as I discuss, in the final chapter, alternative visions of GIS.)

If this seems a loose definition, I would respond that once one begins, as I do, to look at the people who use the technology, the products they use, where they use them, the relations among the technologies, and the legal and moral systems that guide them, one can find no real justification for the use of a more restricted definition.

There is a third way, related to the last, in which one can see this work as a broader analysis. It is broader just to the extent that it seldom operates at the level of the particular. If I occasionally mention one software package or another, one system or another, I do so only to make the argument more clear, and not to suggest that what I have to say is restricted to that particular case. In fact, the issues about which I write, issues of scientific practice and representation and intellectual property and privacy, are issues that are very much general to the technology. The issue of privacy arises in a vector system in much the same way as it does in a raster system. The issue of responsibility for data arises in a distributed system in much the same way that it does in a centralized system. The issue of the relationship between the GIS expert and the subject of research arises as much in a private as in a public system.

If there are disadvantages to this approach, there is a single overriding advantage: The issues that I discuss are long-standing; the issues of representation and privacy and intellectual property and the nature of expertise, social and cultural and moral issues, were around long before there was such a thing as a geographic information system and will be around long after the current flurry of activity is over. To be too specific, to focus directly—even if more precisely—on the details and defects of a particular system is to doom oneself to early irrelevance. I have hoped, here, to offer a way of looking at the systems that will in some way endure.

This brings me to a final point. There are two extremely common and equally dysfunctional ways in which those involved in the production and use of technological (especially technological, but also social or religious) systems respond to works such as this, works by a relative outsider, and works that attempt to look at the systems in which they are involved. The first is to argue that as an outsider one has no right to comment on a system; the second is to counter any general statement with a particular statement, to counter any statement that refers to a particular with a reference to a particular not mentioned. If these are just the strategies invoked by the person whom I mentioned at the outset, they are at the same time strategies that one often sees, in discussions of geographic information systems and elsewhere.

I said that these strategies are dysfunctional, and that is exactly what I meant. The first, the rejection of the outsider, is certainly widely held. Yet it is as certainly rejected by those who claim to hold it. Who, after all, would insist that only artists can say anything substantive about art, only cancer patients can say anything substantive about cancer? What geographer would insist that only the residents of Redding can say anything substantive about Redding? Indeed, those who make such claims are doubly inconsistent, just to the extent that they at once argue for the rationalization of means of data analysis, and make an argument that is in the end an *ad hominem*: "Your argument is incorrect because you're not one of us."

In a sense, of course, the second strategy is just a variation of the first, at least to the extent that it appeals to the up-to-date knowledge of the specialist. But it is dysfunctional for a second, different reason—it responds to an assertion with one of no demonstrated relevance. The simple piling on of evidence is not, after all, really an argument, but merely another rhetorical technique.

Still, these two rhetorical strategies persist as responses to those who comment about geographic information systems, and persist in interesting ways. While in computer science more broadly there has long been an active intellectual engagement by practitioners—for example, by members of the Computer Professionals for Social Responsibility and at annual meetings such as Computers, Freedom, and Privacy—there has only recently developed the beginnings of a parallel in geography or geographic information systems. This is a matter, too, that needs to be explained, and this volume is, in part, an explanation of that fact.

Budapest
December 1997

ACKNOWLEDGMENTS

Earlier versions of portions of this work were presented at Clark University, Edinburgh University, Harvard University, Hunter College, Princeton University, the Rochester Institute of Technology, San Diego State University, the State University of New York at Buffalo, the University of California, Santa Barbara, the University of California, Los Angeles, the University of Washington, and the University of Wisconsin and at conferences sponsored by the Association of American Geographers, the Institute of Electrical and Electronic Engineering Society on Social Implications of Technology, the International Association of Mass Communication Research, the National Academy of Sciences, the National Center for Geographic Information and Analysis, and the Southern California Conference on Technology, Employment, and Community. My thanks to those institutions and to the audiences there for support and comments.

I have received support, financial and otherwise, from the National Center for Geographic Information and Analysis (and through them, the National Science Foundation). The NCGIA's meeting at Friday Harbor, Washington was a watershed. Before, those of us who were already working on this subject were voices in the wilderness; afterwards we were part of a larger community. That remarkable meeting went a long way towards turning the study of the social, cultural, and ethical issues raised by geographic information systems into a "real" area of inquiry. At the same time, it brought together—face to face—practitioners of GIS and critics, in what turned out to be an amiable and spirited exchange. Thanks to Tom Poiker for his work in making the conference happen.

This work has also benefited from discussions at the NCGIA's Initiative 16 meeting on Law and Information Policy for Spatial Databases, the Initiative 21 meeting on Formal Models of Common-Sense Worlds, and, of course, the Initiative 19 meeting on the Representation of Society and Nature in Geographic Information Systems, and at the meeting of Project Varenius.

In writing this work I have received financial support from the Academic Senate of the University of California, Los Angeles.

Portions of this work were carried out while I was in residence at Edinburgh University and at the Center for the Critical Analysis of Contemporary Cultures at Rutgers University.

Special thanks go to the Program on Information Resource Policy at Harvard University—to the staff for their support, and to Majid Tehranian, David Goodman, and especially Tony Oettinger for encouragement and intellectual stimulation. As I finished the manuscript I was saddened by the death of my friend there, Anne Wells Branscomb.

For invaluable support, I'd also like to give special thanks to David Mark, Eric Sheppard and Mike Goodchild. Thanks also to Nick Entrikin, Bill Clark, John Pickles, Bob McMaster, Phil Agre, Roger Clarke, Leah Lievrouw, Sheri Alpert, Helen Nissenbaum, David Flaherty, Charles Raab, Gary Marx, Rohan Samarajiva, Bill Vitek, Wade Robison, Helen Couclelis, Tom Conley, Pat McHaffie, Stuart Aikin, Ken Hillis, John Krygier, Dalia Varenka, and Harlan Onsrud.

Thanks to Claire Barnes, Joan Hackeling, and Travis Longcore for research assistance.

Finally, in a different way, thanks to Mark, David, Gerald, and Melissa.

And, as always, to Joanna.

Portions of this volume were previously published in the following: Michael R. Curry "Image, practice, and the hidden impacts of geographic information systems", in *Progress in Human Geography* 18, no. 4 (1994): 441–59; Michael R. Curry "On the inevitability of ethical inconsistency in geographic information systems", in *Ground Truth: The Social Implications of Geographic Information Systems*, edited by John Pickles, New York: Guilford Press, 1994; Michael R. Curry "Rethinking rights and responsibilities in geographic information systems: Beyond the power of the image," *Cartography and Geographic Information Systems* 22, no. 1 (1995): 58–69; Michael R. Curry "Data protection and intellectual property: Information systems and the Americanization of the new Europe", in *Environment and Planning A* 28 (1996): 891–908; Michael R. Curry "Space and place in geographical decision making," in *Proceedings of the Annual Conference of the Institute of Electrical and Electronic Engineering, Society on Social Implications of Technology*, Princeton, NJ, 1996; Michael R. Curry "Digital people, digital places: Rethinking privacy in a world of geographic information", in *Ethics and behavior* 7, no. 3 (1997): 253–63; Michael R. Curry "The digital individual and the private realm", in *Annals, Association of American Geographers* 87, no. 4 (1997): 681–99; and Michael R. Curry "Rethinking privacy in a geocoded world," in *Geographical information systems: Principles and applications*, edited by Paul A. Longley, Michael F. Goodchild, David J. Maguire and David W. Rhind, New York: Wiley, 1998. The author thanks the publishers for granting permission to excerpt from these works.

INTRODUCTION

Only a few years ago geographic information systems were virtually unknown. They were obscure technological systems requiring large computers, producing execrable output and interesting only to a small few. Today fishing magazines run stories about the relative merits of raster and vector mapping. Software that produces maps but that can also create three-dimensional representations of terrain is a mass-market item. Hand-held global positioning systems are stocking stuffers. Geodemographic systems are routinely used to describe clusters of ten or fifteen households, and their producers are beginning to promise retailers "rooftop geocoding."

So geographic information systems have entered the mainstream. They have done so as part of the burgeoning telecommunications industry, an industry that promises to be *the* area of economic growth in the next decades. And if the telecommunications industry is an area of economic growth, it is demonstrably an area of cultural, social, and legal change. Magazines such as *Wired* celebrate the cultural impacts of the new technologies. And even more visibly, the media are filled with self-referential accounts of the spread of wireless communications, electronic mail, and of course the Internet. Some proclaim a utopian future, of more community, and of new forms of space and place; others, more dystopian, see these technologies as offering havens for pornographers, pedophiles, and terrorists.

And so, we seem to be in a situation in which telecommunications technologies are widely believed to have substantial geographic implications, and a situation in which geographical technologies—geographical information systems, global positioning systems, remote surveillance systems, and automated cartography—are everywhere to be found. Yet geographers have been remarkably silent about these developments. Granted, there has been the occasional piece about economic geography and telecommunications, but these have remained marginal to the discipline of geography. About the only visible manifestation of an interest in the subject is seen in the routine invocation of David Harvey's term

1

"space–time compression." But even there, the wide gap between Harvey's merely suggestive term and the ever-present reality of telecommunication technologies should be telling, as should the failure of geographers to have recognized and followed up on precursors to Harvey's concept, as in Don Janelle's 1969 "space–time convergence."[1] And if geographers seem to have been relatively uninterested in the geographical implications of telecommunications technologies, they seem to have been equally uninterested in the implications of their own, geographical technologies. Indeed, it has only been recently that the great majority of those who use the technologies have not greeted the study of their implications with suspicion and even hostility.

Why this should be the case is not at all clear. In the case of telecommunications technologies, the lack of interest no doubt has a number of sources—the "ineffability" of wireless communications networks, the pragmatic empiricism of many geographers, the inertia of subdisciplinary boundaries. One wonders, though, whether many geographers may not harbor fears that in the end, some critics are right, and that these new technologies will lead to the death of space and place, and hence of their own discipline. Better, on this view, to keep the head in the sand.

And why have geographers not embraced the study of geographical technology? Surely one reason is that they have adopted the view, so widespread, that all technologies are neutral and natural. According to this conventional wisdom geographic information systems comprise a set of tools, neutral pieces of technology, there waiting to be used or misused. One might well study their inner workings, in order to understand where they may go wrong, and where they may be made inefficient, but to study their use, their role in society, is to look to features of the systems that are only contingent, and that are therefore not worth studying. If others have used the system badly, what does this have to do with me? Indeed, to study the systems is to deny that neutrality; to place them under scrutiny is to open the possibility that others, those less knowledgeable, more ideological, will place them under attack.

This double reluctance, to study the geographical implications of telecommunications technologies and to study the implications of geographical technologies, has been enhanced by a particular sort of envisioning on the part of the geographers. On the one hand, geographic information systems are viewed quite narrowly, and much that might be counted as a GIS is not. Many would declare geodemographic systems beyond the pale; others would exclude automated cartography. Some would exclude geographic information systems when they are used for non-geographical uses. Others would exclude facilities mapping systems, or CAD systems used for GIS, or the sorts of mapping functions included in spreadsheet programs and so on. On the other hand, geographers have had a tendency to view the systems, even narrowly defined, as transformative, as the geographical technology par excellence, the technology that will transform the discipline.

Unfortunately, this envisioning process has meant that geographers have both turned away from related geographical technologies, seeing them as simplistic or impure, and from non-geographical technologies, such as electronic mail, the fax, the Internet, and word processing. And this is unfortunate, because there can be no doubt that those technologies have been extraordinarily important to the transformation of academic and intellectual practices, both within geography and without. This double envisioning has blinded geographers to the real technological changes that have fundamentally altered the practice of geography, and of all academic disciplines.

In this volume I shall attempt in a small way to remedy some of this inattention. I shall not, unsurprisingly, manage to lay out an exhaustive economic/political/social/cultural geography of the world in an age of telecommunications technology; nor shall I comprehensively examine every facet of geographical information systems and technologies.

My aims are more modest. They are to sketch out the lineaments of an account of the nature of geographic information systems and technologies, and to suggest some of the ways in which those systems fit within a larger social and cultural landscape.

The volume consists of three parts. In Part I, I ask the following question: What can one represent using a geographical information system? Some might argue that when they are integrated into the most sophisticated systems—for example, for the representation of terrain in three or four dimensions—they are capable of representing literally anything that can be imagined. Is this really true?

I shall argue that there are very substantial limitations to what can be represented using geographic information systems. Putting the matter most generally, my argument focuses on the difficulty that the systems have in representing everyday practices. We live in a world that is constituted of everyday practices, and of the contexts—place, family, community—within which they make sense. It is a world in which one writes software in a computer lab, carries a gun on a hunting trip, and screams with abandon at a basketball game. To do the wrong thing at the wrong place is to risk a variety of consequences; at the least, it is to risk being misunderstood. Indeed, whether in learning a language, writing a program, or having a chat, we demonstrate that we know what sorts of statements and actions, what sort of practices belong where.

But when we look at the representations of everyday practices produced using geographic information systems, for various reasons we have difficulty seeing what is represented *as* practices. And this is in part related to the difficulty we have in making sense of the relationship between the creator and the systems and their output. The systems, at best, are ambiguous about those relationships, and even about the fact that the systems *have* creators.

As we shall see in this section, this problem, this limitation, is complex. It is in part related to the ways in which reason and language are conceptualized within the systems. Both are imagined in highly spatialized terms; language is seen as a set of concepts and relationships that "map" onto the world, and the operation of reason is understood in terms of truth and tables and the like. Further, the representation of space appears at best ambivalent. We find within the systems a routine and unexamined process of "switching," where conceptualizations of objects in space and of their relationships with one another and with that space are subject to an unwitting alteration, one that may only become obvious once its implications are followed out into other areas.

Central to the practice of science have been two processes, which Bruno Latour has referred to as inscription and the production of optical consistency, where inscription is the process of writing—of journals entries, articles, and the like—and to produce optical consistency is to create a set of representations of the world that people are prepared to accept as "going together." In the final chapter of the first section we shall find that in the case of geographic information systems, the nature of inscription is unclear, and the desire for optical consistency has been met by an appeal to a set of technologies of location.

At the same time, implicit in these technologies has been a further appeal, to a view in which the world consists, in the end, of information, and in which the relationships between people, and among people and nature are all, in the end, matters of information. This, though, is not a possible way of looking at the world; it is an image of the world that lacks just what one needs in order to *have* a world, places and the everyday practices that constitute those places, practices such as using language and doing science. In the end, then, while geographic information systems strike one as quite general in their outlook, we need to recognize that they can only create coherent, meaningful output when we recognize them as fundamentally local.

In Part II I turn from the limits of representation using geographic information systems to the question of the relationship between the systems and the institutions within which they operate. One finds in the popular literature, and in much written by those who developed the systems, a common way of looking at developments in both science and technology. On that view, which I term the conventional history, both develop in terms of a kind of internal logic, where the move is from more error to less, or from technologies that are less efficient and capable to ones that are more so. And on that view, there is little reason to study the development of technological systems, other than to satisfy one's craving for immortality—or gossip.

But as we shall see in this section, matters are not at all that simple. Indeed, we shall see in Chapters 5 and 6 that the introduction of geographic information systems into science is associated with a fundamental undercutting of the ability of those using the systems to appeal to what many see as a core set of scientific values, of universalism, disinterestedness, and skepticism. Universalism, the belief that one may achieve very general results, is undercut just to the extent that the systems rely on fundamentally local technologies, where the size and complexities of software, hardware, and data sets make it impossible to specify fully the similarities and differences among systems. Disinterestedness, the operation of science within a set of institutions that are putatively indifferent to the results that are achieved, is undercut by the increasing porosity of the walls of the academy. Here in geographic practice, as before elsewhere, we see a flood of commodities insinuating themselves into the practice of science. Classes are taught in the use of software products, and the instructors are certified by the producers of the systems. Products are provided at low cost, with the assumption that those who learn using hardware and software will later recommend the purchase of these same products. Indeed, commodities are becoming increasingly visible in the academy. As a result, it becomes increasingly difficult to assert believably that social-scientific research, and particularly that which concerns the very industries and institutions being studied, meet the standard of disinterestedness. However honest and well-intentioned may be those doing research using the systems, that research remains for structural reasons suspect.

Finally, scientists are typically seen as skeptical, just as the system of science itself is perceived through institutions like peer review itself to be skeptical. Yet the development of geographic information systems, which has seemed in the ideal case to imply the development of a universal map, a means of combining local systems into one overarching system, has in the process been led away from this ideal. This is because fundamental to the combining of systems is believed to be the development of sets of standards, that will define data formats, and guarantee a consistency of terminology. Yet the standard-setting process and the use of standards, in a different way, move the practice of science, even to the extent of questions of what exists, to the offices of corporations and government. As the use of geographic information systems becomes more general, as it diffuses into government, the number of stakeholders increases and the stakes become greater. What exists becomes increasingly, and with increasing visibility, a political matter.

In Chapters 6 and 7 I turn, at least in part, away from the academy and to the use of geographic information systems in the world. In Chapter 6 we find that the size and complexity of the systems—of software, hardware, and data—have begun to recast the traditional understanding of the task of the scientist, the technician, and the provider of tools. Indeed, it is increasingly difficult to discern who has rights to products created with the systems, and who has the responsibilities. There developed, over the last two centuries, a set of institutions for the definition and regulation of such intellectual property, and those institutions defined these rights and responsibilities. Yet today, with software, hardware, and data increasingly produced within corporate environments and then leased to users, and with those users themselves ambiguously defined, the practice of science using geographic information systems and information systems more generally has come to involve the negotiation with corporate providers, who see themselves as having little stake in the perpetuation of the traditional intellectual property system. Indeed, as we shall see in the case of Europe, the very sorts of data used by geographers are seen increasingly as vital elements of the international system of trade.

In Chapter 7 I turn to the related issue of data protection and privacy. For as data have become more mobile, data about individuals and households have become more valuable. And central to the increasing value of the data has been the ability, using geographic information systems and geodemographic systems, to attach locational attributes to individuals and households, and thereby to concatenate information into larger, more powerful profiles. If in the 1970s privacy advocates worried about the ability of governments using mainframe computers to combine databases using individual identifiers, like the social security number, as common keys, the development of sophisticated means of geocoding has obviated the need for such numbers, and in so doing undercut the most powerful of traditional means for privacy protection.

At the same time the efflorescence of profiles of individuals and households, through commercial geodemographic systems and especially the direct marketing industry, has meant that traditional understandings of the need for a right to privacy, which focused on the damage caused by the dissemination of inaccurate information, no longer capture a central motivation for the development of such a right. That motivation was the belief that people need a kind of metaphorical "space," within which they are free to think and act, to develop personal, moral, and political beliefs without outside interference. But geodemographic profiling leads to a proliferation of individual profiles to which may be imputed all manner of moral and political beliefs. If the individual still retains a measure of ability to think and act "in private," the systems increasingly mean that the fruits of those thoughts and actions exist only within a broader context of imputed thoughts and actions.

In Part III, I turn away from what is, and inquire into a set of normative questions. Most centrally, when using a geographic information system, how does one decide whether some set of actions are acceptable? What are the moral limits of the use of the systems? As should be clear from the previous discussion—from the consideration of the representational limits of the systems, and of their interrelations with social and legal institutions—this is not a simple question. Indeed, one is inclined here simply to advert to some standard of "normally accepted practice."

Yet with technologies as potentially far-reaching as these, such a standard will surely not do; local practice within an office is far too centered on the daily exigencies of producing a product and keeping a job to do justice to the broader implications of a single use of such a technology.

In Chapters 8 and 9 I address these questions from two rather different perspectives. In Chapter 8 I consider several of the ways in which one *might* think about how to act ethically using geographic information systems. Central to most such ways is the appeal to sets of rules. One imagines a system of morals to be a system of rules, and imagines that a person acting morally is one who has acted in accordance with the appropriate set of rules. Yet, as we see in Chapter 1, there are some puzzles when one tries to make sense of how this following of rules operates in practice. And making matters more puzzling still is that this way of thinking about thinking is so firmly a part of the conceptualization of the individual and the mind in the modern era, so firmly a part of the engagement with the world that has brought us the computer and the division of labor. Indeed, as we shall find, moral discourse, in the end, refers to moral practice, and moral practice is a matter of how people do things at particular times, and particular places. The question of ethics cannot be separated from the questions raised by geographers. And to ask how one ought to act is to ask how one ought to act in a particular place—the computer room, California, the university, America. If this seems in the end to offer little in the way of guidance about how one ought to measure one's actions when using geographic information systems, this might well be a sign that it is time to think about the systems themselves.

In that vein, in Chapter 9 I consider some recent calls for a new geographic information system, a "GIS$_2$," to replace the current "PaleoGIS." Some, and particularly those interested in public participation, have begun to suggest in tentative ways that it might be possible to recast the systems, to do so in ways that will make the ethical responsibilities of their users more clear, while, for example, minimizing the possibility of using the systems in large-scale surveillance systems. There, though, we find, paradoxically, that an underlying set of beliefs about just what a GIS can be has from the outset doomed these attempts.

Indeed, this very belief, a belief in technology and space and the mind and the individual and the nature of geographical inquiry, has had another paradoxical effect. It has blinded geographers to the immense geographical impact of a wide range of information technologies, and to the impact of those technologies on geographical practice itself, just as it has focused attention on a talismanic image, of space tamed.

Part I

THE WORLD ACCORDING TO GEOGRAPHIC INFORMATION SYSTEMS

1

REASON AND LANGUAGE IN GEOGRAPHIC INFORMATION SYSTEMS

Geographic information systems seem extremely, and increasingly, powerful. If early systems produced output that was remarkably crude—just to get a map out of a system was often a chore, and what looked like a "computer-generated" map was sometimes a product of as much cutting and pasting as any traditional map— newer systems are easily able to produce traditional maps, spatial analyses and visualizations; even animated fly-overs are a few mouse clicks away. And the rapid move from representation and analysis to simulation and visualization makes it seem reasonable to believe that if existing geographic information systems cannot do everything—and more—that one could do with conventional systems, they soon will be able to. If some have argued that with a geographic information system one cannot represent what one could using traditional means of geographical representation, on this evidence it appears that one can do far more.

Yet I shall argue that, to the contrary, there are very substantial limits to what can be represented with a geographic information system. These limitations fall into three categories. First, the systems incorporate conceptions of language and reason that are limited. These conceptions are important in one way to the extent that the objects that are represented are thereby represented in limited ways; they are important in another just because those creating these objects operate according to a set of images of reason and language that obscure what is being done in the creation and use of the systems. Second, the systems are unable to represent space in ways as powerful as is done elsewhere. And third, the systems have sought representational coherence through appeal to a set of technologies of location, and those technologies have failed in two ways: They have been decidedly local in their scope, and they have at the same time driven a view in which all—place, time, and nature—is simply information.

Granted, it *would* be possible to develop an alternative form of geographic information systems, what some have termed a "GIS$_2$," that could avoid these limitations. But as I shall argue in Part II, the conceptual limitations of the systems are deeply embedded within institutional structures and practices that make such changes difficult to carry out.

In what follows I shall begin with the issues of language and reason. I shall then turn to the representation of space within geographic information systems. And finally, I shall discuss some of the limitations, within the systems, on the representation of people and places. As we shall see, though, the three sets of images are intermingled; indeed, misunderstandings of the nature of place and community are right at the heart of misunderstandings of language, reason and space.

On language

A central issue in geographic information systems is that of language, and a central limitation on the representational capacity of geographic information systems is the way in which those who construct and use the systems conceptualize language. I am not being arbitrary in my suggestion that language is a central issue; we shall see in the later discussion of the Spatial Data Transfer Standard (FIPS 173)[1] that within that standard the establishment of a lexicon, a list of approved terms, has been a central task, as in Table 1.1.

This example points to one central feature of FIPS 173, and of the way in which people commonly think about language. We tend strongly to believe that the words that we utter refer directly to objects or events or processes in the world. If I say "shaft" I refer to "a long narrow passage sunk in the earth." Similarly, when I say "Africa" I very likely believe that there is some place in the world to which the word refers. If I say "diffusion" or "evolution" I likely believe the same, this time about a process. It is almost as though the words contain built-in pointers, pointers that tell us how to find that to which they refer. Indeed, it is quite typical to think of language as in the first instance a set of signs, its primary function to impart information about the world; a set of marks that does not provide information does not even count as a language.

Granted, there are variations in language, and in some kinds of languages this process of signification seems more clear and more formal, while in some it seems less clear and less formal. For example, most people would agree that in the case of "natural" languages, like English and French, rules are more complex and subject to interpretation, and that in "artificial" languages like FORTRAN or

Table 1.1 Items in the world

Shaft	A long narrow passage sunk in the earth
Product	The item or substance produced through an industrial process

LISP the rules are more formal and clear, and more easily applied without resort to interpretation.

If geographic information systems have been used in attempts to model natural languages, the ideal of language appealed to in the *creation* of geographic information systems appears to fall squarely in favor of the artificial. The systems are created within the realm of science and computer technology and within these realms, with their idealization of equations and models, the creation of a language that is a completely formalized set of terms and relationships has long been an ideal.[2]

Although there have been a wide range of understandings of just how such a completely formalized system might work, and especially about the relationship between terms and the world, there has been general agreement that a formalized language must associate sets of characteristics with elements, that an atom needs to be characterized in terms of a certain set of features, a city in terms of another set, and so on. Such a language, that is to say, has been seen as fundamentally typological and essentialist; its task has been seen as one of dividing up the world, and doing so in terms of categories that grasp not just accidental features, but real and important ones.

Now, this image of language and of the possibility of a perfect, artificial language has been central to thinking about science—and has remained so well after the revolutions in the philosophy of science of the last thirty years.[3] It has appeared particularly appropriate to the case of information systems, and of geographic information systems; this is because central to such systems are databases, consisting of items and attributes, where location may be either an attribute of an item, or an item on which attributes are predicated.

This is just what we see in the example from the Spatial Data Transfer Standard; there language is viewed as consisting of proper names, like "The Brooklyn Bridge," types of categories, like "bridge," traits, like "suspension," and relationships, like "connect" and "separate." A language is built up from these elements, and in a fundamental sense we can view language as a *structure* made up of these elements, attributes, and relationships, a structure whose job is to *represent* the world.

One thing striking to a geographer is the way in which this process of representation in language resembles the process that relates a map to the world. In a map we find place names that refer to places, lines for rivers and highways that refer to real rivers and highways, and representations of areas. It is almost as though one could hold the map up to the world—from space—and everything would match. And we find in a certain view of language much the same image. Indeed, one of the most powerful recent variants of this view was developed by Ludwig Wittgenstein, whose 1922 *Tractatus Logico-Philosophicus* elaborated a "picture theory" of language that very much adopts this mapping imagery.[4]

Wittgenstein, in fact, claimed to have hit upon the theory after seeing a court-room recreation of an automobile accident, where models and a map stood proxy for, and represented, the real accident.[5]

So when we begin to explore the relationship between language and geographic information systems, we are treading ground that has already been well trodden. Indeed, if this picture theory was articulated early in this century by Wittgenstein, it is important not to see it as just a kind of flash in the pan. Rather, it embodies views of language and knowledge that are very old indeed. For example, one aspect of this view is the belief that categories like "bridge" or "human" are real, that the words refer to physical or conceptual objects that persist beyond the individuals to which they refer—and that, in fact, can be referred to in any of a number of natural and artificial languages. This realist view can be traced back to Plato's famous views in the *Theaetetus* and *Sophist*, and more popularly to his famous parable of the cave in the *Republic*.[6]

Similarly, we find this view of language laid out explicitly in Augustine's *Confessions*, where it is claimed that

> I noticed that people would name some object and then turn towards whatever it was that they had named. I watched them and understood that the sound they made when they wanted to indicate the particular thing was the name they gave to it.[7]

Here, that is, we learn language by ostension, by pointing.

More recently, as early as the seventeenth century, we see the elaboration of this view in modernist philosophers and scientists like Descartes and Locke.[8] In these last we find the roots of what has come to be an increasingly intense attempt to develop a scientific language which abjures the particular: Western science has since the seventeenth century tried to get rid of the particular, of names of individuals, and to devote itself to the project of creating a set of explanations that apply quite generally. In the ideal case, it has been hoped, one would be able to refer to anything, a lake, a mountain, or a celestial event, in terms of characteristics—including location, time and duration as well as types of processes and materials—alone.

Now, in what I have just said I have spoken of language and geographic information systems in two rather different realms. On the one hand, where I have spoken of the Spatial Data Transfer Standard, I have referred to the ways in which language has been conceptualized *within* geographic information systems. That is, I have referred to the ways in which those who create and use the systems have imagined that language can and ought to operate within the systems. Here, I should add, we need to understand that this is a very general standard. If spectacular visualizations and fly-overs seem to have transcended it, underneath them lies this very conception of language; the landscapes simulated and visualized have, after all, been constructed out of elements contained in the SDTS dictionary.

On the other hand, the construction of a geographic information system that incorporates this conception of language is very often done by people who believe that this conception of language is the same one that they as humans, or at least scientists, use in their everyday life. Just as in a geographic information system "shaft" refers to "A long narrow passage sunk in the earth," in the real world a pie is a pie.

This is a compelling way of thinking about language, both within geographic information systems and without. And as we shall see in the next two chapters, it is a way that both supports and is supported by conceptions of space and place and of individual and community. But unfortunately it is, in the end, quite wrong. It is wrong in a number of ways, but putting the matter most generally, it fails both the empirical test—language does not work that way—and the test of possibility—it could not work in that way.

On rules

In order to understand why this is the case it will be necessary to turn to a connected issue, of the nature of rules, reason and rationality. As we shall see, these subjects are related not only to the question of the nature of language, but also to the question of the nature of computers; hence, the issue of rules and rationality is related to geographic information systems in two very important ways.

Fundamental to the modernist image of language as a map or picture is the appeal to a particular notion of "rule." There, and in geographic information systems as in artificial languages, rules are seen as rigidly and unambiguously determining the outcomes of the operations within which they are involved. If a rule states that

$$a + b = b + a$$

then it is assumed that this rule will operate for any value of 'a' and 'b', and that once the values of the variables are entered, there is no need for individual judgment. At the same time, violations of a rule, as when an inappropriate data type is used, lead to predictable outcomes, also built into the rule. Finally, it is typically assumed that it must be possible to make explicit the relation between a rule and its application.

If this view appears in geographic information systems and in science more generally, it also extends to mathematics. And in that way, too, it connects to geographic information systems—particularly if we take seriously Russell and

Whitehead's argument that mathematics can be wholly derived from logic.[9] For there computers are merely complex logic machines, ones that work in terms of a two-valued logic where every question may be answered yes or no, true or false. In fact, just these two states—here I omit reference to the far less important analog computers—can be represented by the positions of an electrical switch, where on = true, off = false.

And so, especially from the point of view of a person who sees the role of science as the creation of discursive accounts of the world, accounts of the form "This x is a y," or "x is in relation b with y," the computer seems a perfect tool, and one that has no important limitations. It can engage in any form of reasoning at all, and in doing so can provide a basis for the language of science.

Moreover, when we turn our attention to geographic information systems we seem to find nothing to make us begin to doubt. After all, in a raster system the GIS need contain only statements derived from ones of the form like "At (x,y) at time t_1 there is an 'f' with characteristic 'k'."

In fact, we need not even be interested in science itself to find computer reason and human reason congenial companions. It is a commonplace to imagine that we think in terms of sets of rules, and in all manner of situations. From "In the US always tip a waiter fifteen percent" to "In Britain always drive on the left and overtake on the right" to "Always show up to work on time," we appear to be people guided by rules. And this appears, moreover, not to be simply a Western predilection; anthropologists like Levi-Strauss and linguists like Chomsky have argued that these rules are in some way built into the human mind.[10]

In the case of geographic information systems this appeal to rules is especially evident in attempts to develop artificial-intelligence-based systems, such as the ones used in human way-finding.[11] Such systems do not presume simply to be lists of the paths that people have taken; in even a small area the number of such paths would be impossibly large. Rather, they embody a practical syllogism, a set of rules for getting around. The rules are of the form "If you want x you ought to do y." In this way they attempt in a more elegant fashion to constitute a universal system, one which can deal not just with trips that people *have* taken, but also with ones that they *might* take.

So in several ways we find embodied in geographic information systems notions of reason which appear commonsensical and consistent one with another. They are notions of reason as involving a kind of binary thinking, and thinking which operates in terms of rules, whether in the traditional "A or not A" and "If A then B" of logic, of the "a + b = c" of mathematics, or the "If you want A you ought to do B" of the practical syllogism. This image of reason appears to be very general indeed, to characterize human reason without leaving a residue. In this way it is an image that suggests that it is quite possible to imagine a computer that can think.

16

As is well known, this view has been formalized in the famous Turing test, where the test of whether a machine can think is whether one can distinguish its output from that of a human.[12] And in fact, I would argue that, in some cases, using a geographic information system one can indeed produce results that by any reasonable criterion must be seen as indistinguishable from the products of human reason.

Take, for example, the placement of names on maps. As any student of cartography knows, this is a task that may be easy or may be devilishly difficult. The basic goal is fairly easy to articulate; one wants to place names so that they are clearly connected with the features with which they are associated, so that the more important ones predominate, and so that all are legible. In placing names by hand one usually starts with some rules of thumb. Place the more important ones first, and then work through the less important. When points surround a central point, place the names away from, and not toward, the center. And so on. In learning manual cartography one learns just such rules. And in one's first attempts these rules are often in mind. But gradually any cartographer develops a set of personal strategies, personal preferences, about where to start—upper left?—and how to proceed—in patches, or straight through? Ultimately, one is no longer even aware of the rules. One just chugs along, placing name after name, and resolving problems on the basis of what "seems best."

Now, in spite of this, computers can indeed be taught to place names on maps, and if they are not now perfect there seems to be no reason why they might not at some point become so. Programs for placing names use artificial intelligence, where the program consists of a set of rules that may be gradually altered, in order to work better and better. The computer can even be told that in certain circumstances—in highway maps, for example—it will be acceptable to move the location of a town in order that the name be rendered readable.

What is essential here is that it is at least in principle possible to use a computer to locate place names on a map in such a way that no person could tell that those names had not been placed by a person. One might, indeed, create a test, such that an interrogator points at a location on the map and an invisible cartographer places a name beside it. This is, of course, a version of the Turing test of computer intelligence, and many would agree that if, in fact, the interrogator is unable to determine whether the names are being placed by a person or a machine there is no reason not to attribute intelligence to the machine that did it. Here, then, is a situation where it seems highly plausible to claim that machine reason and human reason are indistinguishable. Here the use of a geographic information system appears to create no new restrictions on what can be represented.

Note, though, that this is a *local* example. In accepting it, as I think a reasonable person must, one is in no way committed thereby to accepting the

17

proposition that with respect to language and rationality geographic information systems impose *no* limits. Indeed, one needs to see that one source of the difficulty in accepting the notion that there are local areas within which representation is unfettered, while there are other areas of great limitation, is the computer itself. For the computer seems itself so strongly to represent an image of reason as a universal phenomenon. Similarly, the map, and now geographic information systems, seems equally to present clear and compelling images of the nature of language. In both cases the most common images simply turn our attention from what we must really do, which is to explore case-by-case the limitations of geographic information systems in action, neither assuming that the sleek images that they conjure guarantee their utility nor presuming that the systems are, because of the failure of those images, in every case fatally flawed.

What started out as a simple question, then, has turned out to be quite complex. At the outset the question of the ways in which the use of geographic information systems might implicate us in notions of language and reason that are constrained, that do not allow us adequately to represent the world, seemed as though it might have a simple answer. For some, certainly, the answer is simply that a geographic information system uses only emaciated notions of language and reason, and so of course must be unable to represent what we want. For others, the answer is quite the opposite; the systems are extraordinarily general, able to take on the most abstract notions of reason and language, space and time, nature and culture. With a geographic information system we can represent the findings of scientists, and thereby all that is important to knowledge.

I have suggested, though, that matters are far more complex than that. Both the advocate and the critic have been taken in by an image, an image of rationality and language which shares a great deal with the image of the machine. It is an image of the system, of the whole that can be decomposed into its parts. It is an image of a system that can be seen *as* an image, and that can be seen, often, on the image of the map. This image is at play in the usual conceptualization of language and its relationship with the world, of reason, and of technology itself. But once we actually look at the operation of language, or reason, or technology we find that these images do us little good.

Beyond the traditional view

I suggested earlier that in the case of language there turn out to be some difficulties when one attempts to put into practice the modernist image of language as a map, and the same is true in the matter of reason. Of course, although we operate every day as though the image of rule-following as a clear-cut mechanical act is the correct image—and more importantly, one to which we advert in our own rule-following—we know very well that the following of rules by humans is seldom that simple. The problem is, though, that there is a strong tendency to see those complexities as irrelevant—when they in fact are right at the heart of the matter.

And indeed, in this matter the computer provides especially good examples. Some programs respond to a particular situation with the statement (of a rule) "Press any key to continue." But in fact, if you strike <Shift>, <Control>, or <Alt>, nothing at all will happen. The "real" rule is "Press any key except <Shift>, <Control>, or <Alt> to continue."

Now, it may seem as though this is a trivial example. After all, the three exceptions are quickly learned; one quickly learns the "real" rule. But this is just the point, that in order to get on one learns to *do* something. But one learns not so much to apply a new and more complex articulation of the rule as just to act in a certain way. One learns that certain ways of acting work. In fact, one of the great difficulties that new users have in approaching computers arises from their assumption that the "rules" written in technical manuals are to be taken literally, to be obeyed to the letter. Many who have been involved in writing such manuals, and especially earlier ones, just took for granted a great many practices, which they assumed that everyone knew, but which were in fact the outgrowths of long periods of training and socialization. Questions as obvious as the identity or non-identity of keys labeled <Return> and <Enter>, or of 'A' and 'a' were often omitted from such manuals, since it was assumed that everyone knew the answer.

Here one might be tempted to argue that this problem arises because computers are so new. And this is absolutely true, but once again supports the point that I am making. Although I have not made a comprehensive study, I would venture that very few official driving manuals begin with assertions such as "In New Jersey you may or may not drive with your automobile air conditioner turned on." This just seems like something that everyone is expected to know, in the same way that *not* everyone is expected to know what to do when arriving at an intersection where the rightmost lane is a bicycle path. There the situation is too new, too rare; it needs to be spelled out, and it is.

It might, though, be thought that this failure on the part of rules to fully spell out their consequences is specific to areas outside of science, to areas wherein people are just not very careful about their ways of thinking and their use of language. But this would also be a mistake. For in fact, in the laboratory we find that not enough is spelled out in manuals and articles to allow us to replicate experiments; if it were there would be no notion that one person is more skilled than another in the lab, and we would not be at all worried at the prospect of being operated on by a surgeon who had learned the craft through a correspondence course.

Indeed, even in the case of mathematics do we find that rules are not totally spelled out and, more important, that they *cannot* be. In his *Philosophical Investigations*, Wittgenstein gave up on the modernist theory of language, the one that has become so much a part of common sense, where language is seen as mapping onto the world. The story is that Wittgenstein saw the futility of his Tractarian view during a confrontation with Sraffa.

> One day . . . when Wittgenstein was insisting that a proposition and that which it describes must have the same 'logical form', the same 'logical multiplicity', Sraffa made a gesture, familiar to the Neapolitan as meaning something like disgust or contempt, of brushing the under- neath of his chin with an outward sweep of the fingertips of one hand. And he asked: 'What is the logical form of that?' Sraffa's example produced in Wittgenstein the feeling that there was an absurdity in the insistence that a proposition and what it describes must have the same 'form'. This broke the hold on him of the conception that a proposition must literally be a 'picture' of the reality it describes.[13]

In beginning to sketch out an alternative, he offered the following example. Imagine that I ask you to take the following series, "2, 4, 6, 8, . . . ," and go on with it. You continue, "10, 12, . . . , 1996, 1998, 2000, 2004, 2008." I stop you, "Wait, that's not what I meant; I meant you to do "1998, 2000, 2002, 2004, ..." Now, I may after the fact be able to say that I had "in mind" that you continue "in this way." But surely, Wittgenstein argued, when I originally asked you to continue I did not have "in mind" every number from 2 to infinity. Rather, in our culture we learn a certain way of "going on," a certain mathematical practice. It is *not* that the entire series is somehow built into the rule, but rather that when we say that someone has learned the rule we mean that we are confident that that person will engage in a set of practices that are the ones in which we believe that we would engage. As he put it,

> To obey a rule, to make a report, to give an order, to play a game of chess, are customs (uses, institutions).
> To understand a sentence means to understand a language. To under- stand a language means to be a master of a technique.[14]

Here again, to think that the entire series is somehow built into the rule is to be bewitched by an image.[15]

It might be argued here that in the case of geographic information systems some have moved away from the association of language with rigid and explicit forms of rules. For example, some involved in artificial intelligence appear to attend to certain practical forms of knowledge, including Gilbert Ryle's "knowing how" and Michael Polanyi's "tacit knowledge."[16] Similarly, arguments have been made that under some conditions it may be desirable to appeal to what appears a "looser" form of rule following, that of fuzzy logic.[17] Yet if this development in one sense does move away from the rigid image of "language as a map of the world" that has guided so much thinking about science, it does so by moving it back one step, for the application of fuzzy logic is seen as rigidly rule-bound as any other. Indeed, it is a commonplace in this literature that this tacit knowledge can be encoded in the same way as can explicit knowledge, and that this can be done in the cases of both declarative and procedural knowledge.[18]

The difficulties to which Wittgenstein pointed—difficulties in formalizing the operation of language—have been addressed in the case of geographic information systems by David Mark. Mark has noted that the systems must regularly fight to connect their technical terms with less rigorously defined geographical terms, and that this connection is made far more difficult both by the operation of any individual language and by the inconsistency of languages.[19] If it is hard enough to make sense of the English use of the term "lake," it is more difficult to develop means for accurately translating from English to French even in as simple a case as this. In fact, Mark's work suggests on the basis of empirical investigation in geography what others have discovered elsewhere, that the language of science is not so easily formalized, and that the traditional subject-predicate notion of language has little to do with the actual use of language.

But here Mark's work suggests the need for deeper changes, as it suggests the possibility that within some of the areas of science which seem most rigorous there is operating an unexpectedly ragged notion of language. His solution refers to the work of George Lakoff, where the reference of terms is set in terms of exemplars, which may be concrete examples or abstractions, and where terms are connected more by family resemblance than by essence.[20] This more recent view clearly operates within a tradition begun as long ago as 1957, in studies of artificial intelligence, where Newell, Shaw, and Simon argued in favor of an appeal to what they termed "heuristics," as an alternative to the more rigid rules of formal logic.[21] Here the suggestion was that we ought to look at human thought not so much as a matter of the application of explicit rules but rather as a matter involving the appeal to images or exemplars.[22] It seems to me that this appeal to the power of images in the matter of language and rules is quite right. Indeed, I shall argue in Chapter 2 that geographers have been strongly influenced by a belief in certain images of space, and that those images of space have had an impact not only on thinking about space itself, but also on thinking about language and reason.

Nonetheless, this appeal to heuristics still fails to develop adequately an understanding that the following of a rule is in the case of any language, natural or artificial, a type of *practice*. That is, the application of rules is *never* built into the rule, and neither is it built into the mind of the person applying the rule. From the statement of a rule itself it is not possible to determine how it would in every case be applied; moreover, there is nothing in the rule itself which will tell us when we can feel justified in asserting that the person who appears to be following it is really doing so in the sense that we believe that it should be followed.[23]

Indeed, the work of a number of philosophers, beginning with Wittgenstein, has suggested that the central problem with the modernist view of rules and language is that it fails to see that rules and language operate only within linguistic communities. Using language and following rules are practices, matters of learning and habit. As such, according to Wittgenstein, they can be understood only against the background of a set of social practices. This is the basis of his well-known assertion that "If a lion could talk, we could not understand him."[24] For it is not just that the lion would be speaking a language that we do not understand; rather, every aspect of the everyday life of the lion is so foreign that we would not know where to start. By contrast, the idealization of explicitness tends to suggest that it is possible to imagine knowledge as a matter of a relationship between a body of knowledge and an individual, and therefore as a process to which communities and groups are irrelevant.

And it is important to understand that this feature of rule-following is general, that it applies in every case. It applies, that is, in science and in geography. So the notion that the need in learning a GIS is to move beyond the written work and learn by doing does not make the systems different from science; rather it shows that in this regard they are *just* like science. Indeed, it makes no sense to imagine it possible to divorce the products of science from the processes through which they were created, just because those judging those products can understand them only to the extent that they recognize that those processes were fundamentally matters of practice.

Here again, the case of placing names on maps turns out to be a good example. I earlier suggested that there is strong reason to believe that it is possible for a computer to lay out names on a map in a way that cannot be distinguished from the way it would be done by a human. In this case, I concluded, we are justified in claiming that there is no real limitation imposed through the use of the computer; computer and human seem equally to be "using reason."

I chose this example for two reasons. For one, it is a clear case of success. The other reason speaks directly—and even more centrally—to the issues raised in this chapter, and especially to the matters discussed in the last section. It is a commonplace that there are two sorts of work, that which fundamentally uses intelligence and that which does not. Administration and management are often seen as work which requires intelligence, craft work as a sort that does not. Indeed, the recent claims that educational systems need to be revamped in order that graduates better meet the needs of a changing economy simply support this view of work.

On this way of looking at things, the cartographic designer is using reason or intelligence, while the technician, who just carries out orders, is not; rather, he or she is simply engaging in routines. Here the designer "knows that." The technician "knows how." In fact, on this view, locating place names on a map is not, when people do it, a matter of the application of intelligence at all; it is a craft, rather like painting a window or installing a kitchen cabinet.

It is a tribute to the power of an image, and of the institutions that underlie it, that this view has managed to retain its credibility, for anyone who has been engaged in production cartography knows first that that work requires a seemingly unending flow of decisions, and second, that in a given workplace with a stable set of designers and technicians, the design work always gradually devolves to the technicians, as the designers come to be more and more managers. Here, vastly different job titles mask similar work.

But more important, we have seen that the similarities are even greater, that activities like computation and research which *look* like applications of reasoning very much like that which we associate with a computer are themselves better seen as human practices. This, finally, raises the issue to which I shall next turn, the issue of space. For the ability using a geographic information system to represent human practices rests fundamentally on the ability to invoke adequate notions of space and place.

2

ON SPACE IN GEOGRAPHIC
INFORMATION SYSTEMS

I suggested in Chapter 1 that there is a connection between the way that space is conceptualized in geographic information systems and the ways in which language and reason are conceptualized. In part this was because the conceptualization of language typically defers to a particular image of space. It turns out, though, that right at the outset the case of space in geographic information systems is more complicated than that of language. This is in part because in geographic information systems multiple conceptions of space are routinely at play. If within geographic information systems homage is commonly paid to certain conceptions of space, the representations created using the systems very often appeal to other ways of thinking. Moreover, in everyday practice the users of the systems typically involve themselves in yet others.

I hasten to add that this is not an unusual state of affairs; we find it all across the sciences, and beyond. Yet in the case of geographic information systems, where space seems so important, it will be equally important to spell out just what is going on.

On the nature of space

There are a variety of ways of "dividing the pie" in discussing the nature of space.[1] But of particular relevance here is the following. We can lay out conceptions of space on a continuum, where at one end are those wherein the relationship among objects in space is strictly contingent, and where at the other are conceptions where the objects in space have very strong, even necessary and intrinsic relations with one another and with the space in which they are located.

Consider a waste bin placed on the pavement. As people walk by they toss things in. And at the end of the day it is quite full, of pieces of gum, candy wrappers, cigarette butts, newspapers, soft-drink cans and so on. It does not seem far-fetched to say that the objects in the bin have nothing in common other than proximity, and that they have no real relationship with one another.

Putting it a bit differently, the bin is still the same bin whether it is empty or full.

Mundane as this example is, it is very much the way that we typically think of space. Notwithstanding contemporary physics, we conventionally view celestial space as a big void, full of stuff, where the void would be the same if all of the stuff magically disappeared. Similarly, we typically imagine that nation states and neighborhoods and factory buildings are very much like the waste bin and the void; empty them out, and the container remains. If the examples that I have given are mundane, we can find ones that are less so. Perhaps most obviously, we often think of geometrical space as a real space, where the points are characteriz-able in terms of some metric, and the points are simply locations in that previously existing space.

This container view has a long and not-so-simple history. On the one hand, we find it in the Greek atomists Leucippus (fifth century BC) and Democritus (c.460–370 BC), and in the popularization of their work in the later Lucretius (c.96 BC – c.AD 55).[2] There the universe is conceptualized as a large void, with matter consisting of individual atoms. But we find a similar view in Aristotle (384–322 BC).[3] In some respects his understanding of space was very different from that of the atomists, but he too articulated a view in which space could be seen as a container. Any place, he believed, is defined in terms of the inner boundary of its containing medium, and so we can see that medium as a kind of container.

And perhaps most notably, this is the view that we see articulated in classical physics, as in Newton:

> Absolute space, in its own nature, without relation to anything external, remains always similar and immovable. Relative space is some movable dimension or measure of the absolute spaces. . . . Absolute and relative space are the same figure and magnitude, but they do not always remain numerically the same. For if the earth, for instance, moves, a space of our air, which relatively and in respect of the earth remains always the same, will at one time be one part of the absolute space into which the air passes; at another time it will be another part of the same, and so, absolutely understood, it will be continually changed.[4]

If this view seems commonsensical, it remains that there are others, which—at least after a bit of examination—seem equally commonsensical. Consider the following examples: I'm at a friend's house, cleaning up after a party. I put a dish away, and he says, "That doesn't go there." I'm at a party, where everyone is polite and sedate, and someone gets drunk and starts being belligerent. My reaction, "What's he doing here; he doesn't belong here." I grade a student's paper, and note on it, "This sentence seems out of place." A family with several young children moves into a retirement community, provoking the response: "Children don't belong here." I could continue, with examples referring to women, African Americans, Jews and so on.

In each of these cases appeal is made to the idea of "belonging." And in each case the relationship between the place and the object or action or person is very different from the relationship that we saw in the Newtonian, absolutist view. For here that relationship is not merely accidental or contingent; rather, it is a matter of necessity. For the dish or the drunk or the children to be in the wrong place is to disrupt the order of things.

Like the absolutist view this is a common one, and like it, its use extends beyond common sense, to the arena of science. Indeed, this very view was articulated by Aristotle, who argued in the *Physics* that a place is the innermost surface of the object bounding the place. And who, moreover, argued there that we can understand the motions of objects simply by seeing that objects have their own natural places, and that solid objects fall because they belong, with their kin, in the earth.

But if this Aristotelian way of looking at the world is a common one, we need to see, too, that there are other seemingly different ways of characterizing space that share this view, that there are intrinsic relationships between the objects in space and space itself. For example, in the early eighteenth century, Leibniz articulated an alternative, that saw space not as an absolute, but rather as the consequence of the interrelationships among objects.

Indeed, if we turn to the famous debate held between Leibniz and Newton's stand-in, Samuel Clarke, or to Leibniz's own *Monadology*, we see Leibniz developing a view in which the universe consists of elements that are intrinsically connected one with another. For Leibniz it made no sense to talk about a space with nothing in it; space, in fact, only came into existence with the objects that we now think of as filling it. There we find an understanding of the nature of the universe that seems the very opposite of that of Newton.[5]

As Leibniz put the matter, people develop a conception of space when

> They consider that many things exist at once and they observe in them a certain order of co-existence, according to which the relation of one thing to another is more or less simple. . . . When it happens that one of those co-existent things changes its relation to a multitude of others, which do not change their relation among themselves . . . we then say, it is come into the place of the former; and this change we call a motion in that body. . . .
>
> And supposing, or feigning, that among those co-existents, there is a sufficient number of them, which have undergone no change; then we

may say, that those which have such a relation to those fixed existents, as others had to them before, have now the *same place* which those others had. And that which comprehends all those places, is called *space*.[6]

Hence, for Leibniz space does not exist as a separate entity, a container, but rather exists only in the relationships among objects; Newton—and we—are led to think that space exists prior to the objects in it by what amounts to mental sloth. And because space is for Leibniz constituted by the objects in the universe, there is a necessary relationship among those objects and that space.

Space known and represented

If Leibniz's analysis is interesting because of the way in which it characterizes the relationship between objects and space, it is interesting for another reason. It points to the ways in which we can gain knowledge of space. In fact, he argues, we never really *see* space, we can never have direct knowledge of it. Rather, the best that we can do is infer its existence from the things that we really do see, namely, objects.

Yet as Kant showed, there is a problem here, and it turns out to bring us right into the thick of issues raised in cartography and geographic information systems. The problem is this: If we only attribute spatiality to the universe after having become acquainted with the objects that make it up, where do we get that idea of space? We certainly do not, on Leibniz's own admission, find it in the world. But if it isn't there, where is it? Kant's answer was that we attribute spatiality to the world because the concept of space is built into us. We, as knowing subjects, order the world, attributing to objects and events in it not only spatiality, but also temporality and causality.

Note, though, that in making this argument Kant is in one way allying himself with Leibniz, just to the extent that he is arguing that with respect to space we find some element of necessity. Here, though, the necessity resides in the relationship between the knower and the known, the person and the world.

And indeed, just as we can distinguish accounts of space in terms of the way in which they characterize the relationships between objects and space itself, so too can we distinguish them according to the ways in which they characterize the relationship between the person and the world. As we have just seen, in Kant that relationship is strong indeed, as it is in Leibniz, while for Newton it is in a sense quite irrelevant.

Now, if we turn away from those who theorize about space, we find that in those who have represented the world there are, explicitly or implicitly, just such conceptions of the relationship between the knower and the world. Perhaps the most obvious example is in art. In the early fifteenth century Filippo Brunelleschi (c.1377–1446) developed a mathematical system for the representation of the world. That system, which is really the fountainhead of modern systems of perspective, placed the eye of the viewer at a single point. By virtue of the establishment of a series of lines, it appeared to place the objects within a painting in such a way that they looked "real."[7]

When we look at Brunelleschi's system (Figure 2.1) we have a tendency to imagine that what he has done is develop a scheme for representing Newtonian space on a canvas. He has looked at the world and, like Newton, seen it as consisting of an empty container-like space, within which objects are arrayed. Yet there is an important difference between the two. For where Newton has imagined that, paradoxically, we can describe this container in objective terms that seem to suggest that we are outside of it, Brunelleschi has established a system wherein the viewer is right at the center of things. Indeed, his system privileges the position of the viewer.

As Martin Kemp described Brunelleschi's first work using the system,

> The painting of the Baptistery was executed on a wooden panel which was probably a square with sides a little less than 30 cm or one foot in length. . . .
>
> Having painted the vivid patterns of the inlaid marble of the Baptistery in such a way that 'no miniaturist could have done better',

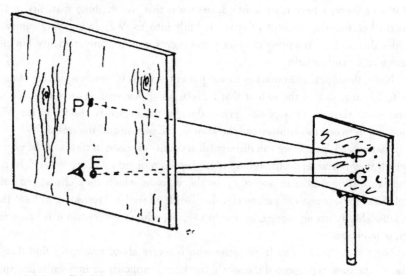

Figure 2.1 Brunelleschi's system

Source: Martin Kemp (1990: 13).

Brunelleschi constructed a form of peepshow to heighten its illusion. He drilled a small hole in the panel at a point equivalent to that at which his line of sight had struck the Baptistery along a perpendicular axis. The spectator was required to peer through this hole from the back of the panel at a mirror held in such a way as to reflect the painted surface.[8]

It is worthwhile to contrast this image of the relationship between the viewer and the viewed with alternatives, and two are of substantial importance. The first is the system for the representation of space embodied in earlier, medieval paintings. The system used in the Middle Ages for the representation of objects in space often strikes the contemporary viewer as peculiar. People and things vary in size; a painting often appears to have within it several scenes, which may involve the same people, or may involve events at different times; and the entire piece seems, typically, quite flat.

Yet the apparent strangeness of these paintings begins to disappear once we begin to see that they were created within and require of the viewer a very different system of representation. Indeed, if paintings created in the modern era using Brunelleschi's system of linear perspective seem to appeal to and support a Newtonian view of space, those of the Middle Ages see the world in a different way. The viewer did not approach the painting with the expectation that the painting would somehow array things in a grid-like space. Rather, the painting was meant to depict scenes of social and often religious significance, where significance was a key. More important figures were larger, or were placed in the foreground; related and supporting figures were placed elsewhere. To view the painting was not simply to place oneself at a peephole. Rather, it was to engage the stories that the painting represented. As Gombrich put it,

> The medieval artist, like the child, relies on the minimum schema needed to "make" a house, a tree, a boat that can function in the narrative.[9]

In such works one's location as a viewer was quite irrelevant, for what defined one's *real* location was one's social and religious status. And in that sense, the painting was a mirror of the world, and one of a world in which what was important was the question of where one, intrinsically, belonged.

If we turn to a second case we find a very different relationship between the viewer and the work. In the second century AD, Ptolemy devised a means for the representation of the surface of the earth on a flat surface. The problem that he thereby solved, he argued, was that of how to do geography. Geography, he argued, is the representation, by a map, of the portion of the earth known to us, together with its general features. Geography differs from chorography in that chorography concerns itself exclusively with particular regions and describes each separately, representing practically everything of the lands in question. . . . It is the task of geography, on the other hand, to present the known world as one and continuous.[10]

But if it appears that Ptolemy's method (Figure 2.2), of the imposition of a grid over the earth, is Newtonian in spirit, Alpers has suggested that his motivation and that of those who followed was quite different.

Whereas Albertian perspective posits a viewer at a certain distance looking through a framed window at a putative substitute world, Ptolemy and distance point perspective conceived of the picture as a flat working surface, unframed, on which the world is inscribed. The difference is a matter not of geometry—in basic respects effectively the same—but of pictorial conception.[11]

Continuing, she notes that

On these accounts the Ptolemaic grid, indeed cartographic grids in general, must be distinguished from, not confused with, the perspectival grid. The projection is, one might say, viewed from nowhere.[12]

Here is a view that we can truly see as consistent with the Newtonian, absolutist impulse.

We have seen—so far—three very different ways of representing objects in space. They might be characterized as in Table 2.1. And this leads us right into the problem of representation faced in geographic information systems.

A conventional distinction in geographic information systems is between those that are raster-based and those that are vector-based. Though the furor that informed earlier debates about the relative merits of each has now subsided, it remains that the two systems are likely themselves to remain, in a sort of uneasy truce. They are likely to remain in part because each seems so ideally suited to certain functions, but also because each seems in a fundamental way to appeal to and support one of the theories of space that I mentioned earlier.

On the one hand, in a raster system we imagine the world as a giant plane—a round one, but a plane nonetheless. We approach the world by seeing everything on this plane as decomposable into isolable entities of a determinate size. In a sense, when we create a system we begin with a pre-established space and fill it up, much as Newton believed that objects came to fill the universe. This sort of

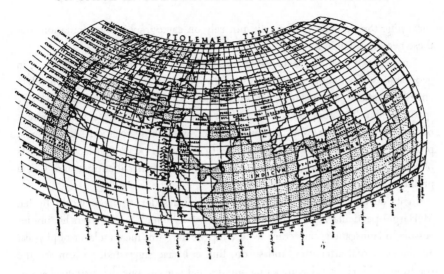

Figure 2.2 Ptolemy's projection, as reproduced in the sixteenth century
Source: John Noble Wilford (1981: 27).

system often involves the creation of several layers. The making of decisions, then, about industrial location or highway corridors or direct mail, is a matter of analyzing the relationships among the layers, through an overlay process.

This view of space seems very much a matter of common sense. The earliest full-scale systems were, after all, land information systems, designed for the representation in cartographic form of a world laced by lines of latitude and longitude. And so it made sense to adopt a view of space that appeared to derive directly from long-established practices and institutions within cartography and

Table 2.1 Conceptions of space

Conception of space	Representation of space	Example	Viewer
Aristotelian	Narrative and symbolic	Medieval	Viewer a participant
Newtonian	Linear perspective	Brunelleschi, Alberti	Viewer in uniquely privileged position
	Distance point perspective	Ptolemy	View from nowhere

surveying. (Some early systems did not rely on cartographic representation, but these need not detain us here.)

Further, this way of looking at space seems almost a matter of definition; in geographic information systems, by convention, data are established with reference to locations, and again by convention, that location is specified in terms of a spatial metric. The adopted spatial grid forms a structure, and the actual content of the system consists of attributes associated with those locations. In the most primitive systems a fact 'a' is associated with a location (x,y); that fact 'a' consists of a statement 'Fa', where 'a' is a name, category, or other identifier and 'F' is a predicate or attribute.

In its current form this approach in some ways derives from works by Ian McHarg—particularly his influential *Design with Nature*[13]—and from his predecessors in landscape architecture, where for McHarg this meant a layer of physical attributes, visual attributes, and so on. But as I have suggested, we can see the tradition of McHarg as having a long intellectual lineage, one that extends back as far as Greek atomists, but one that in a more familiar form derives from the era of the rise of modernism.

Yet we might wonder here what has become of the Leibnizian conception of space. In an important sense, it has re-emerged in vector-based geographic information systems. There one sees a representation of the world not as a composition of overlays on a pre-existing planar surface, but rather as a surface that is itself constructed by the arrangement of objects. And there we can see the process of representation as one of creating an image that is itself structurally homologous with the world itself.

Indeed, this is just the view that we find in designers such as Christopher Alexander.[14] For Alexander, the design process is a matter of looking at the objects that one wants in the world, and then seeing their possible, probable, and necessary interrelationships. The world emerges from this pattern of interrelationships. And we see it in mathematics in works by R.H. Atkin, and in related geographical work by Peter Gould.[15] At the same time—and this connects us back to the issue of language—we see it early in this century in the work of the early Wittgenstein. There he argued that

> What any picture, of whatever form, must have in common with reality,
> in order to be able to depict it—correctly or incorrectly—in any way at
> all, is logical form, i.e. the form of reality.[16]

> A picture has logico-pictorial form in common with what it depicts.[17]

And

A picture represents a possible situation in logical space.[18]

Here, finally, we might wish to believe that the matter is settled. We have found in contemporary discourse about objects in space an extraordinarily complex set of conceptions and of systems of representation. There are conceptions that see objects and space as tightly, even intrinsically connected. And there are ones that see these relationships as loose indeed. There are conceptions in which the viewer is intrinsically connected with the viewed, and ones in which that relationship is far more tenuous. And there are conceptions in which the viewer has a position of privilege, just as there are ones in which the viewer is in a far less enviable position.

We have also seen that these are views that are commonly held, outside of geography and the use of geographic information systems, and that they have long histories. That having been established, we might very well feel that the issue of the nature and limits of representation in geographic information systems has been settled. If there are differences among the systems one can simply develop rules for translating from one to another.

On switching

But matters are not that simple. And in fact, a series of processes and strategies operating within the practices and institutions of geographic information systems guarantee this complexity. The first I shall call switching. Consider a classic block diagram (Figure 2.3). In this case, clearly, the illustrator wished the diagram to be viewed from a particular position. Indeed, Lobeck's treatise is full of examples of just how to create such an illusion.

By contrast, consider Figure 2.4, a map of the same landscape. Here we see a map created using a projection—in this case and at this scale, which projection is hard to determine—where it appears that the viewer is simply "above" the landscape. The viewer, in Alpers' terms, is viewing the landscape "from nowhere."

Now, the difficulty, with respect to the conceptualization of space, should be clear. If we start from the position that we see in the original diagram and pivot ourselves up and over the landscape, at some point a "switch" occurs. At some point we are no longer viewing the landscape from a single point; we are viewing it from nowhere.

This phenomenon has been studied in Gestalt psychology, and Norwood Russell Hanson used the concept as a means of making sense of certain phenomena in the history of science.[19] Yet there is a critical difference between the phenomenon that Hanson described and the form of switching that I am describing here.

Consider a classic example of switching in psychology, and one that Hanson

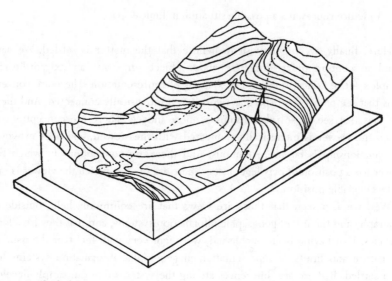

Figure 2.3 Block diagram
Source: Armin Kohl Lobeck (1958: 143).

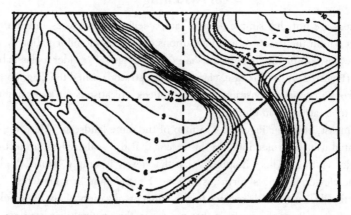

Figure 2.4 Map
Source: Armin Kohl Lobeck (1958: 143).

uses (Figure 2.5). We can see this picture first as an old woman and then as a young woman, and the process of seeing one and then seeing the other is indeed a process of switching. Yet when we compare this process to the one that I have just described there is a difference. For here it is always clear that I am seeing the image as one or another. But the process of switching at work in cartography and geographic information systems is not at all that simple or obvious. If we refer back to Figure 2.3 and Figure 2.4, when does the switch in conceptions of space occur? It is difficult indeed to say, and we may only become aware of having made the switch when decisions made on the assumption that we were working within one conception or representation of space lead us astray.

If this process of switching can concern the location of the viewer, it can also concern the nature of the relationship between the viewer and the viewed. For example, in the case of the block diagram in Figure 2.3, it makes sense to imagine the viewer *as* a viewer, and as standing outside of the scene. But when the "same" conception of space is invoked to create a fly-over, the viewer is no longer a viewer in quite the same way. Few, after all, would describe the experience of being in a flight simulator—or at least a competent flight simulator—as a matter of viewing a scene. One might better see it as a matter of the "representational barrage" that John Krygier describes as typifying the representations created in the Great Reconnaissance of the American West.[20] As Alexander von Humboldt put it,

> Panoramas are more productive of effect than scenic decorations since the spectator, inclosed as it were, within a magical circle, and wholly

Figure 2.5 Switching

Source: Norwood Russell Hanson (1958: 11).

removed from all the disturbing influences of reality, may the more easily fancy that he is actually surrounded by a foreign scene.[21]

There the viewer, as has so often been intended, is likely to characterize the experience as direct, and unmediated.

A third form of switching, related to the first two, occurs as one moves from more literal to more metaphorical conceptions of spatiality. Here consider the case of numerical taxonomy. It is conventional there to characterize the procedure as the application of the traditional distance formula

$$d^2 = a^2 + b^2 + \ldots + n^2$$

In the case of a two or three dimensional version of the formula we usually have no difficulty translating the algebraic expression into a geometrical figure. But when, as in geodemographic models, we begin to use a formula where the nth term is the 250th, matters get considerably stickier. Most of us know few people who can really "draw" a four-dimensional figure, let alone a 255-dimensional one. And in fact, it is clear that at some point the use of the term "dimension" takes on a new and purely metaphoric meaning. Yet where is that? If I develop a three- or four-dimensional terrain model, I can treat that model as a piece of algebra or one of geometry. And if it is a piece of geometry, I can move from a wire-frame version to one that is fully rendered. And I can then place that model in a system within which it becomes a flight simulator.

At some point there I have switched, moved from a purely conceptual relationship with the model, one where I can imagine it as a conceptual object outside of me, to one that has a visual existence, and to one that projects around me that representational barrage that seems to transform it, and my relationship with it in yet another way.

There is a fourth and final form of switching. This one derives from the situatedness and intertextuality of representations. Consider the map in Figure 2.6. To say that our understanding of it is situated is simply to say that it matters whether this is a map appearing in a brochure for people searching for jobs, a kit for someone attempting to find markets for drafting supplies, an introductory cartography textbook, or—as it indeed did—a book on postmodern geographies. This may seem a simple, even a painfully obvious point, one known by any beginning cartography student. Yet once we see that in Soja's work there is what amounts to a call to understand the entire spatial organization of society in what amount to Leibnizian, relational terms, the matter may not be so simple. At the same time, to say that the reading of a map—or a model—involves intertextuality is simply to say that important to the situatedness of the map or model are other maps or models that the creator and the reader have brought to the table.

In Chapter 3 I shall discuss three additional processes that come into play as we use the products of geographic information systems. Each of these in a different way affects the ways in which the systems are used, and the limitations on those uses. In closing here, though, it is useful to remind ourselves of what we have seen, and what remains to be seen. In the practice of science, in using technologies, and in our everyday lives we routinely appeal to conceptions of space, and we do so in ways that are sometimes explicit, sometimes implicit. For the most part these conceptions are longstanding; several can be seen as having roots as far back as classical Greece.

At the same time, just as we conceive of space in various ways, so too do we represent it in various ways. And whether in conception or representation, we tend to imagine that we are, variously, intrinsically connected to the objects of the world and to a spatial system, or separable and separate from them.

These are not idle issues. In obvious ways they are connected to the very conceptual bases of geographical explanations, and if that seems an overstatement one would be advised to return to the battles in physics in the seventeenth and eighteenth centuries, between Aristotelians and Newtonians, and Newtonians and Leibnizians. It is clear in those disputes just what is at stake.

But there is another issue here, and that is the way in which the various

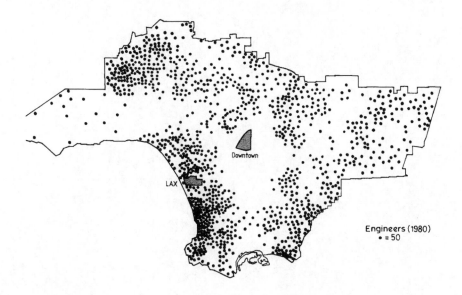

Figure 2.6 Engineers
Source: Edward Soja (1989: 211).

conceptions and representations come to be marshaled in support of social and political goals. For Aristotle's view, where everything has its place, can be—and is—invoked in support of the status quo, just as Newton's universe-as-container has provided support for neo-classical economics, where people turn out in the end to be just as mobile as atoms in a void.

3

OPTICAL CONSISTENCY, TECHNOLOGIES OF LOCATION, AND THE LIMITS OF REPRESENTATION

Bruno Latour has argued for the importance of two facts in the development of modern science. The first is the desire for what he terms "optical consistency," the desire to be able to imagine the world as all of a piece, and to represent it visually in ways that support that consistency. And the second is inscription; he argues that central to the development of science has been the process of writing, where scientists in fact spend increasing amounts of their time writing, and talking, arguing, and writing about writing.[1] And central, too, has been the desire to make those writings, those inscriptions, increasingly easy to comprehend.

Do we find these elements at work in geographic information systems? In a sense, yes. After all, we might want to consider the process of switching that was described in the last chapter as both a cause and a by-product of the use of certain forms of visual representations, where a useful form of representation is one that is able to appear simple, and thus to mask differences among viewers and users. But it seems to me that when we look at geographic information systems and at cartographic representations we find some peculiarities. And it further seems to me that understanding those peculiarities will help us understand the limits of representation using such systems.

Optical consistency and the nature of inscriptions

Consider the matter of optical consistency. Trivially, perhaps, we might think of it in the following way. I purchase a camera, and I then walk around with the camera strapped to my face, seeing everything through the viewfinder. Everything would appear in the viewfinder with systematic distortion, and always from my own, unique point of view.

But as we have seen in the last chapter, in the real world all sorts of things intervene. Images drawn from a particular vantage point meld with those drawn as if from nowhere; the ways we look at images are colored by the relationships of those images to others; and even when we are looking at an image or model that appears to have adopted a view from nowhere, we are—demonstrably—somewhere.

So in fact, this ideal of optical consistency is not and cannot be achieved. Yet at the same time, Latour is surely correct in saying that in their everyday activities scientists both work toward such a consistency and act as though it is attainable and even attained. But consistency itself is not a natural feature of images or models; rather, it is a characteristic that we agree to impute to models and images. In an important sense, consistency is constructed, and in this construction we appeal to a notion of what belongs. Consider the rendering by Le Corbusier of the community at Passac (Figure 3.1). This is a standard architectural rendering, meant to be a "view" of the proposed development. We see buildings, streets, and landscaping. But what is missing here? We see no people. There are no mailboxes, trash containers, trucks, cars, birds, fire hydrants. There are no potholes, electrical and telephone wires. And why not? Because the inclusion of those elements would violate the principles of optical consistency within architectural rendering. There is a similar process at work in Figure 3.2, as there was in traditional bird's-eye views.

For Latour, one of the means for achieving optical consistency is by the use of inscriptions. An inscription can be anything written, though it seems clear that in the case of science these inscriptions need in some way to be public, to be shared. But here, when we turn to geographic information systems, a problem arises. If we look again at Figure 3.2, we note that it is in fact a representation of a

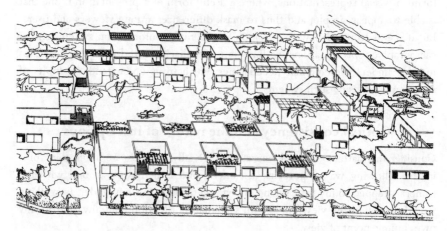

Figure 3.1 Le Corbusier's Bordeaux-Passac

Source: Le Corbusier (1986 (1931): 252).

computer screen. And there is no reason to assume that such a screen has ever been printed onto paper. Moreover, we note that in the case of such terrain visualization models, the viewer is capable of moving her location, of moving around the scene, higher above it, and even through it.

But then, in this case where *is* the inscription? Or to put it in another way, when the users of these systems talk about what they are seeing, what are they talking about?

Or take the following example:

Mirror Worlds?

What are they?
    They are software models of some chunk of reality, some piece of the *real world* going on outside your window. . . . A Mirror World is some huge institution's moving, true-to-life mirror image trapped inside a computer—here you can see and grasp it whole. . . .
    The 'geography' perspective is a natural starting point, sometimes. The picture on your screen represents a real physical layout. . . . Now

Figure 3.2 ARCPLOT

> you see inside a school, courthouse, hospital, or City Hall. . . .
> Eavesdrop on decision-making in progress. . . .
>
> You can use the Mirror World's archival propensities to discover the
> history or background of anything that concerns you. . . . [2]

David Gelernter's image of the computer system of the (near) future, in his *Mirror
worlds: Or the day software puts the universe in a shoebox: How it will happen and what it
will mean*, incited one reader, the "Unabomber," to reply to the author with a
bomb. Yet Gelernter's account of mirror worlds should strike many as very much
in the tenor of what is happening in GIS. So, once again, where is it?

It seems to me that in the case of geographic information systems we need to
look well beyond the ordinary processes of inscription, beyond field notes and
dissertations and monographs and articles and abstracts, beyond grant proposals
and time sheets and purchase orders. Because here the construction of common
sense, the creation of a belief that different systems are optically consistent, is
strongly tied into a set of elements and processes that are in some ways separate
from geographic information systems, and that certainly predate them. I have in
mind what I term the "technologies of location."

The technologies of location

On standards

When I refer to the technologies of location I am not referring simply to the
objects found in the back of a surveyor's truck, and neither do I have in mind all
of the note-taking that surrounds the surveyor's work, although both are surely
important. Rather, I refer particularly to those technologies, and their supporting
institutions, that are central to making location public and general. It is made
public in the sense that access to information and to the means of gathering that
information is in principle free; it is general just to the extent that a system is
adopted within which any location can be characterized *and* within which
conflicting systems can be translated using rules that are themselves general and
public.

Here we need to turn right at the outset to the role of government. The relationship between geography and the state has always been a close one; after all, at
the time of the birth of Christ the Greek geographer Strabo asserted

> the utility of geography—and its utility is manifold, not only as regards
> the activities of statesmen and commanders, but also as regards knowl-
> edge both of the heavens and of things on land and sea, animals, plants,
> fruits, and everything else to be seen in various regions. . . . [3]

Today governments have been keenly involved in funding the development of geographic information systems, for both civilian and military purposes. And in doing so they have promoted the adoption of sets of standards, definitions that are far richer, more complex, and more thoroughgoing in their attempts to be universal than ever before.

In the case of geographic information systems one can actually discern four types of standards being advocated and developed. The first are associated with the creation of a formal system for marking out the face of the earth, and here latitude and longitude form the obvious example. The second are standards for categorizing elements and attributes; here the desire is to create a universally applicable vocabulary, so a river here will be a river there. The third is a system for measuring and formalizing the accuracy of data within a system. And finally, there is an emerging set of standards for the interchange of data. Here I shall focus on the first, and shall return to the others in Chapter 5.

The formal system for marking locations on the face of the earth is, of course, old indeed. We find it in the Greek concept of 'klimata'. We find it elaborated into something that looks rather like our current system of latitude and longitude as long ago as Hipparchus, who lived from c.190–120 BC. And in Ptolemy (AD 90–168) we find a map of the world that lays out with some precision the location of various places (Figure 2.2).

Now, it might very well seem that by the seventeenth century, and the adoption by scientists like Descartes and Newton of a theory of absolute universal space, the matter was quite set; we live on an earth over which is stretched a grid, and that grid is a local version of the larger, universal grid in which the earth floats. Yet matters are not that simple, and for the following reason. If it is true that the earth was long ago well mapped, it is also true that these were strictly *local* maps. Consider the United States, for example. While it is possible to locate any point in the country in terms of latitude and longitude, every point is so-located only in terms of some base point, some standard datum. In this country the North American Datum of 1927, based on a point in Meades Ranch in Kansas, was the standard for many years, and out from it fanned a system of 200,000 secondary points, located between 3 and 25 kilometers apart.[4]

So if this system appears to be a universal one, it in fact is not. Rather, each local survey is only carried out in relation to some local point, and each of those local points is in turn related to the main point at Meades Ranch. In an important sense, then, to the extent that it was based on the North American Datum of 1927, each map of the United States can better be seen not as related to a universal grid, but rather as related to this central point. The system of maps of the world might better be seen in terms of the image of plate tectonics, as a system of sheets wrapped over a globe, where sometimes the edges fit, sometimes there are gaps, and sometimes they meet and buckle.[5]

From the point of view of geographic information systems, a critical change occurred with the replacement of this datum point by the North American Datum of 1983. This new datum was not based, like the old one, on the laying out of a series of traverses from known points. Rather, the new datum promised a new *method* of establishing standards.[6] Here the key has been the use of a set of satellites as the datum points. Based on them, the United States government has developed global positioning systems which rely on the Doppler shift in radio signals from a series of satellites, and which are able to provide accurate real-time information about location.

Here the standard for location on the earth is no longer the surface of the earth itself; rather, the standard has now become the earth's center of mass, around which the satellites orbit. And the data from these satellites rely fundamentally on geographic information systems; indeed, without the systems it would be difficult to conceptualize them. Here we see the establishment of a new set of locational standards which is in turn based on a new image, where each map can now be seen as part of a seamless web.

Or do we? The matter is actually rather more complicated than this. Citing the need to develop a system more general than that usually used by local surveyors, but not so general as latitude and longitude, the NAD 1983 adjustment included the means for translation into the State Plane Coordinate System, a system whose boundaries are largely political, and which therefore makes reference to more traditional ways of thinking about location.[7] Indeed, the idea of a seamless sheet anchored to a set of fixed points, and hence of a system that defines space in an absolute way, remains here simply an image. For what has in fact been effected is the translation of one set of relative coordinates into another. After all, the orbits of the satellites are themselves defined through an interaction with the mass of the earth, the sun, the moon and other planets. In the end, one must calibrate the orbits of one's satellites against something taken as absolute, and that something needs for the most pragmatic reasons to be on the face of the earth.

And so, when we look at systems used for determining one's location on the face of the earth, we actually see two things. We see an image, of a seamless web, a grid into which it ought to be possible, in principle, to incorporate all other maps. And at the same time—at least if we look more closely—we see a much messier system, one in which the various sheets are stretched and sometimes torn in order to make them fit together, where it makes more sense to see each sheet as defined primarily in terms of the interrelationships among the objects and locations that are mapped. But in either case, we see technologies of location, objects and procedures and inscriptions that can be invoked in support of the idea that locational information is public and general, and provides an underpinning for the development of geographic information systems within which those inscriptions are far more evanescent.

The geodemographic world

Another of the technologies of location has developed at the intersection of government and industry. It is the geodemographic system. First developed in the 1970s, geodemographics arose out of a fortuitous set of events that began in the previous decade. On the one hand, the United States Bureau of the Census created the GBF-DIME (Geographic Base Files—Dual Independently Map Encoded) system as the basis for the computerization of the decennial census. The project, in effect, transferred to computer previously hand-produced census tract maps for all of the urban areas of the country; it provided a means of determining—more or less—the geographic location for every address in those urban areas. On the other hand, following soon after his election in 1960, President Kennedy ordered a process of rationalization of what is now the United States Postal Service, and this led to the development of the ZIP (Zone Improvement Plan) Code.

Before these developments, the Census and the Postal Service shared the appeal to traditional means of determining who lived where, methods that relied on memory and on local knowledge; but those methods and that knowledge were specific to the organization. After the development of the GBF-DIME files and the ZIP Code it became possible—if not always easy—to link that knowledge.

The ZIP Code was immediately seized upon by the marketing industry. For example, in a 1967 article "Zip Code—New Tool for Marketers," Martin Baier hailed it as "a 'built-in' and universal means of geographic identification."[8] As he put it

> [T]he ZIP Code System offers a new, unique opportunity; the way it has been put together (although devised for quite a different reason, namely postal efficiency) just happens to fit many marketing needs.[9]

And with the addition of the GBF-DIME files, marketers—and others—were able to combine data from the Census Bureau's mapping project with data from the Postal Services Carrier Route Information System (which consisted of a comprehensive listing of mailing addresses) and thus to create lists that provided a geographical location for every address in the nation. Thus was born the foundation for modern geodemographics, as marketers noted that the new system allowed them a much more powerful way of applying an insight that they had long had, that "People with like interests tend to cluster."[10]

By the early 1970s the first of a growing group of corporations were established, all with the aim of applying this insight—that people tend to cluster—in the context of increasingly powerful and affordable computer technology (Table 3.1). These companies, engaging in what they termed "geodemographics," moved rapidly away from the mere collection of information about individuals.[11] Faced with restrictions (such as the American Fair Credit Reporting Act of 1970) on the collection of individual information, they began to take the household as the primary unit.[12] At the same time, relying on the theory that people cluster with others like themselves, they began to develop methods of "data profiling."

In a popular description of geodemographics, Michael Weiss describes the system developed by Claritas, a system very much like that developed by other commercial concerns.[13] Within the systems people who reside within particular areas are presumed to share certain characteristics, which can be summarized as "lifestyles." In the Claritas system there is, for example, a neighborhood named "Blue Blood Estates." It consists of people who subscribe to Barron's at 9.44 times the average for the nation as a whole; who buy Jaguars at 17.58 times that rate; who drink bottled water at 2.54 times that rate; and so on. The systems vary with respect to the number of lifestyles that they take to be "basic," but in general, they divide the United States into forty or so such basic units.

Although newer systems introduce additional statistical complexity, the underlying technique in geodemographics is numerical taxonomy. There a set of data units—such as census tracts or even households—are arrayed in a virtual space where each attribute constitutes a dimension. In this n-dimensional space, households or areal units are considered to be "alike" just to the extent that they are close to one another when the distance is measured using the classic equation:

$$d^2 = a^2 + b^2 + c^2 + \ldots + n^2$$

And so, in a geodemographic system "lifestyles" are the units to which one belongs. But that belonging is strictly a contingent fact. A lifestyle is, after all,

Table 3.1 Some geodemographic systems

System	Vendor
Atlas MarketQuest ™	Strategic Mapping
DNA ™	Metromail: R. R. Donnelley
Lifestyle Selector	National Demographics and Lifestyles
MicroVision ᴿ	Equifax National Decision Systems
Niches ™	Polk Direct
Prizm ᴿ	Claritas NPDC
Solo ᴿ	Trans Union

nothing more than a statistical aggregation, and one can change one's lifestyle just as easily as one's hair color. A quick move, a switch in a few buying habits, and I am suddenly a different person. And in fact, this contingency is built into the very methods of geodemographics.

So within such a system we see an ongoing reconceptualization of the objects that make up the world. Social and cultural groups are redefined as mere aggregations of information about individuals. Places are defined as locations attached to which is information, in the form merely of contingent sets of features or attributes. And cultures and places come to be seen as composed of or inhabited by individuals whose names and bodies come increasingly to be armatures to which are attached geodemographically constructed identities.

The world as information—and beyond

Indeed, if the technologies of location associated with setting standards for measurement on the surface of the earth have provided practical and rhetorical support for the inscriptions at the heart of geographic information systems, the second set of technologies, concerned with the location of people, bring us to the core of geographic information systems, and to the limits of representation within the systems. This is just because the second set of technologies so forcefully support a view of the world as information.

I do not mean that those who use geographic information systems appeal strictly—or even loosely—to a technical definition of information. Rather, I mean something entirely more simple; it is imagined that everything in the world can be characterized in terms of information. It might be thought here that I am proposing a mystical view, one in which there are things that simply cannot be described. But although it may be that there are such things, that is not my claim. Rather, my claim is that much of what happens in the world may be describable after the fact in terms of information. After I return from lunch, or start World War III, I may be able to articulate a set of reasons, in terms that suggest that I was consciously using information.

But in fact, these are very often *post hoc* constructions, and much of what people do in the world they just do. Take, for example, the issue that we saw earlier, of the following of rules. When asked why I interpreted data in a particular way, I very likely appeal to a rule. When asked why I appealed to that rule, I likely appeal to another one, about what constitute good statistical inferences. But as Wittgenstein put it,

> "How am I to obey a rule?" . . . If I have exhausted the justifications I
> have reached bedrock, and my spade is turned. Then I am inclined to say:
> "This is simply what I do."[14]

There reaches a point at which my use of language or of mathematics appears no longer subject to question. At that point I have gone beyond information. "What has to be accepted, the given, is—so one could say—forms of life."[15] And in so doing I very often appeal to what we do in particular places.

On place

In fact, one of the central ways in which people construct a world made up of places—home, the workplace, the nation—is by establishing and maintaining sets of activities or practices.[16] As with social groups, the relationships between people and the places in which they live are often strong and enduring. Indeed, many see those in their families or their hometowns or churches or even their workplaces as so close to them that "they are part of me." No one who has seen a recently widowed person wither away can see this as simply a matter of speaking. At the heart of much patriotism is the idea of the individual as an intrinsic part of a much larger body politic.[17] And as the geographical literature has long made clear, this relationship between people and places is a longstanding one; as David Lowenthal has noted, in the Crusades people died of nostalgia, of homesickness.[18]

Whatever the disadvantages of this idea, it remains that the sense of belonging in some intrinsic way to a larger body, whether a nation or a neighborhood, has long been a critical part of both the individual's motivation to act in some larger interest and of the group's ability to exhort the individual to such actions. And this intuition about the world gets played out in one way in works by people like McHarg and Alexander. The questions that both ask are, in quite fundamental ways, questions about what belongs where, and about how to use a particular process to make sure that we only put in a particular place things and people that truly belong there. But, as in geodemographics, this Aristotelian aim is masked in their work by the rational methodologies that they propound. Those methodologies represent the relationship between the individual and the place or neighborhood as one that is strictly contingent. Put the other way around, a place is a concatenation of individuals, connected through a set of contingent relationships. The place itself is simply a spatial location which has attached to it such a set of individuals.

I noted above that for Aristotle everything in the world belongs somewhere, and where it belongs is an essential part of what it is. Put in other terms, the relationship between an object and where it belongs is not simply fortuitous, or a matter of causal forces, but is rather intrinsic or internal, a matter of what that thing actually is. When things are not where they belong, when they are out of place, they cannot truly be themselves.

This way of thinking is very much common sense; we see it played out every day. Police question people who are not where they belong, who are in a BMW in a poor neighborhood (possible drug buyers) or lingering on a busy street corner (possible prostitutes) or even simply walking where people normally drive (pick a city in southern California). We commonly tell our children that they don't belong in certain places—the liquor cabinet or the neighborhood billiard hall. And we do so because we believe that people of their sort, "good kids," don't frequent those places. Being a good kid does not "go with" being in places such as those.

But what leads us to believe that in a particular case the elements belong together? This is, after all, an enormous question, and one that has long exercised theorists of art and architecture. Here I would point to two ways in which we reach this conclusion. The first is concerned with time. Here it will be useful to look back to the work of a geographer at the turn of the century, Paul Vidal de la Blache. Vidal believed that in certain times and places people develop ways of life, groups of human habits related to tools and machines that are historical accumulations of habit.[19]

Vidal, though, believed that over the course of time certain ways of life undergo a kind of transformation. They develop a sort of "personality." Indeed, he began his *Tableau de la Géographie de la France*, itself the introductory volume of Lavisse's *Histoire de la France*, by arguing, after Michelet, that "France is a person."[20] But—and this later had important implications—he continued,

> A geographical individuality does not result simply from geological and climatic conditions. It is not something delivered complete from the hand of Nature. . . .
>
> [A] country is a storehouse of dormant energies, laid up in germ by Nature but depending for employment upon man. It is man who reveals a country's individuality by moulding it to his own use. He establishes a connection between unrelated features, substituting for the random effects of local circumstances a systematic co-operation of forces. Only then does a country acquire a specific character differentiating it from others, till at length it becomes, as it were, a medal struck in the likeness of a people.[21]

So for Vidal ways of life which have developed in particular places may, over the course of time, become so engrained that the place and the people become almost one; the place acquires a personality. Some of his students, it should be added, argued that in France, Brittany cannot be seen as having had time to develop such a "personality," since it underwent structural changes in the fifth century, and that for a great many, Vidal included, Germany was totally without personality.[22]

On time and narrative

One need not, I think, take this view quite to that extreme, but should, rather, take from Vidal and his students the fundamental insight: that the development of places takes place over time. Indeed, if there is a lesson to be learned from a historical reading of accounts of the creation of the humanized world, right through to the creation of the new virtual places on the Internet, it is the role of time; what varies is, more often than not, the amount of time that is perceived as necessary to the creation of the relationship of belonging, and not the need for time itself.

If time and persistence are so important to the creation of both natural and human places, one might wish to argue that there is a tool at hand for representing time, and that that tool is available to the users of geographic information systems. After all, a number of ways of representing time, used in geography, have been adapted for use in geographic information systems. There is of course the classic view of time as a cube, one seen in Hagerstrand and in Berry's classic representation of the nature of geographical inquiry.[23] There time is seen as a third dimension, with location and theme (or some other variable) making up the other two. This view may envision time simply as a vector, such that any "slice" is arbitrarily taken, or may itself be based upon slices, in what amounts to a more up-to-date version of the classic landscape approach of Jan Broek.[24] Variants of this view have been adopted in geographic information systems, where time is typically imagined as a kind of fourth dimension in a Cartesian framework.[25]

But if this view of time has seemed open and flexible, even universal, this has turned out not to be so. Indeed, it is now a commonplace that lived time, involving projections toward the future which at the same time function without any explicit awareness of chronological time, are not easily accounted for in a traditional four-dimensional system. Variations of this view have made their way into accounts of history and, more recently, of science.[26]

In fact, geographic information systems appear here to have provided a way around this difficulty. That a geographic information system can be used as a platform for audio or video displays, including those which incorporate stories and documentary footage, appears to make real the possibility that the representations within such a system are no more limited than in any other medium, and may even be less limited.

Indeed, we see just this recognition in recent versions of geodemographics. For example, in Polk Direct's Niches™ we see the typical "Working Hard" household (average income under $20,000; average age 49) described in the following way:

> My late husband Jerry and I used to kick ourselves all the time for not going to college. Oh, I'm doing alright [sic], but it's hard. I really have to

work a lot just to stay on my feet. . . . Anyway, even though I work a lot, I do take my share of cigarette breaks. My mom used to smoke too, but what she can't understand is all the health foods I buy. What can I say? One thing my mom can understand though, is the fact we buy only American cars.[27]

Here the appeal to narrative is meant to convey a more vivid sense of the individual as one who has a unified identity, where the parts are intrinsically interconnected.

But if in the case of language a stark image belied a richer practice, here the opposite is the case; an image of richness, of the possibility of narrative complexity and of the representation of place, belies a practice much less rich. The reason has to do with the difference between chronology, which is easily representable within a GIS, and narrative, which is not.

Commonsensically we think of the past as composed of a set of determinate events which are themselves basic and from which a primitive narrative, a sort of ur-narrative, can be constructed. This narrative would result from our combining all of the smaller narrative accounts of the past.[28] Or, alternatively, we can see these smaller narratives as abstracted from that ur-narrative itself. But this possibility, so plausible at first, begins to make less and less sense when scrutinized. Leaving aside the obvious problem that the view reflects a naive sort of Platonism, when we in fact attempt to combine narratives we find that they don't easily fit together. Moreover, it even becomes unclear what the basic elements are: an event in one—the migration of Europeans across the Atlantic in the nineteenth century—is in another something very different, perhaps a whole range of events. Moreover, that larger story can be told in quite different ways, thereby giving a differential importance to the events so abstracted. For example, focusing on individual communities within a region would give migration a role different from that generated by focusing on the position of the region with respect to the changing world industrial economy.

What does this say about our understanding of the past, or even of events separated from us in space rather than time? It suggests that while there may be facts about situations, facts that are in some important senses determinate, at the same time those facts only enter into events through the medium of narrative, and narrative is always constructed by *a narrativist*. Hence it points to the centrality of an author or researcher to the creation of explanations of changes in the landscape.

The problem of the nature of narrative has been the subject of much debate, in history, economics, and geography.[29] Simply put, chronology deals with "clock time," with the regular sequence of events. By contrast, narratives deal with a specific question, what happened next. When put in this way narratives seem similar to chronologies. But the focus of the question, on what happened next, is in a narrative answered in terms of a particular *point of view*. And as Arthur Danto pointed out—and as any reader of Faulkner's *As I Lay Dying* or viewer of Kurosawa's *Rashomon* knows—while it is possible to take a number of chronologies and put them together, when narratives are put together one gets not a single more complete story, but rather a series of juxtapositions.[30] To the extent, then, that a geographic information system involves the linking of a series of narratives, those narratives can only be contingently associated with the larger temporal, and chronological, framework.

By contrast, where one attempts using a geographic information system to create more fully integrated products, as in the case of GIS-based virtual reality systems, a second form of constraint arises. This is because a narrative has a narrator. In the matter of place, Entrikin has drawn upon Nagel to show the way in which attempts to create spatial explanations which have no authors, which like certain maps, are "views from nowhere," inevitably fail, as the elucidation of such explanations ineluctably moves toward a recognition of the point of view implicit in the account.[31] This same logic operates in the case of explanations and accounts involving time. A narrative which appears to be a view from nowhere in fact inevitably embodies a point of view; it can always be seen as the expression of the point of view of a real or virtual author. And this author is in a fundamental sense standing outside the project, and is therefore standing outside the temporal framework created by the project.[32]

In fact, many people today would argue that the relationships he describes are not "real," but rather are imputed by the author. I may say that I believe myself to have a deep and abiding relationship with my birthplace, but I only notice that after the fact; while there—and certainly while being born—I was simply taking the matter a step at a time. As Yi-Fu Tuan has put it, I only develop a "sense of place" once my initial relationship of "rootedness" has been upset, or called into question.[33]

Indeed, the development of necessary relationships is something that occurs only in the interaction between the author and the audience. An author describes and exhibits sets of relationships, with the intention that some be seen as necessary; the reader reads those relationships, sometimes responding to the author's intentions, sometimes being blind to them, and sometimes seeing necessity where the author did not. This is as true of the author and reader of a geography text as it is of the "author" of a reflection on his childhood and the listening dinner guests.

And it is also true of traditional cartography. Consider a traditional paper map. I lay it out on the table, and right in the center there is a blob of jelly. My reaction: "That doesn't belong there." Most of us who have spent time making paper maps have at one time or another had the same reaction when we see, almost hidden, the impression of a piece of rub-on lettering. And when we say that "that doesn't belong" we typically do so by referring to the intentions of the cartographer: "Of course she didn't mean to leave that shred of rub-on lettering hidden next to that label."

Just in this way, the entire process of learning to be a cartographer, to produce maps, is one of learning what belongs where. The cartographer asks: "Where does one put the neat line, the scale, the various labels, and so on?" And the reader of the map routinely makes judgments about those works in terms very much like those used by artists: "That doesn't belong there," and "That goes together nicely."

But does a GIS have a narrator? Or an author? And is reading a GIS like reading a map? If the answer to these questions is "No" then we shall need to conclude that geographic information systems do not have available to them the traditional tools for the representation of necessity, of belonging. And indeed, I would argue that that *is* the case. Some of my reasons will become more clear in Part III, where I discuss the issue of intellectual property. But putting the matter most basically, in order to believe a person's account of the world, whether in a story, a map, or an equation, we need two things. We need to be able to say that the person "meant what she said," and was trustworthy. That is, we need to be able to tell our own story of the author's authorship. Who is the author? How did he or she become an expert on this subject? Where did the data come from? How did the author acquire them? If the author's sources were other people, why would they trust him or her? What sort of process of selection went on? Who was involved in the selection? Does the author have obvious biases? Was the author motivated by greed or malice?

Once we have become familiar with an author and an author's work, we seldom ask these questions. But at the outset, as we learn whether to believe a politician or parent, scientist or minister, these sorts of questions are right at the fore. And the answers to these questions by no means involve mind-reading. Judgments about a person's creativity, trustworthiness, intelligence and honesty are always based on a range of experiences, of the person in a variety of situations. As we shall see in Chapter 7, it was the disruption of the traditional patterns of those situations, in the late nineteenth century, that was at the heart of the drive to codify the right to privacy. And indeed, we see much the same in science, as we look at the history of the professions.[34]

If in order to believe an author we need to have at our disposal the resources on which to base that belief—or at least need to have no good reason to question our habitual belief—in order to take seriously the claims in the work about the relationships among objects and the whole we need something more. We need to feel that we can, really, "read" the work. All of us have been in the position of taking up a new area of study, and feeling that we were completely at a loss when it came to judging the works that we read. And a great many people feel the same way when confronted with twentieth century music, or contemporary art. The reader or the member of the audience feels unable to judge; he feels as though he lacks the knowledge and ability to see behind the work to those things that would make it comprehensible. The author and the work remain a mystery.

Now, when we look at a paper map we feel in many cases that we do know what we need to know in order to read it. We feel, at the very least, that if we find it confusing there are at least a finite number of questions that we might ask, that would lessen our confusion. Further, when we look at a set of data that have been subjected to statistical analysis we find much the same; we believe it possible to disassemble the interweaving strands of data and concepts, and "see" if what has been done makes sense. And that is because we believe—and are trained to believe—that the systems of reference within a map can be specified. Maps, as literary theorists have recently discovered about their own objects of inquiry—are intertextual; they refer to other maps. But we are taught that they do so in an orderly way. Granted, here, that recent work in the history of cartography greatly broadens the range within which we must look for the sources of meaning of a map.[35] But one might argue here that these newer works simply broaden the range of what is relevant, without redefining the process.

Yet in a geographic information system the processes of defining the author, establishing a relationship of trust with that author, and establishing what one takes to be good reasons for feeling competent to judge the system are far more complex. Hardware is, inevitably, a product of multiple creators, and as simple and straightforward as it may seem, a string of hardware problems—concerning, for example, the handling of floating point calculations—suggests that even hardware is not completely transparent to the viewer. Software, by contrast, is notoriously and by general agreement complex and difficult to read. At one and the same time, algorithms are kept secret and even the algorithms that are not secret can in principle never be fully tested. And finally, even under the strictest regime of standards and testing, the belief in the adequacy of a set of data can, in the end, be merely an act of faith. One may here counter that the adoption of standards by data providers (see Chapter 5) allows one to believe in the adequacy of one's data. But of course, one can only do so if one believes the people who set and administer the standards, and they are likely to be even more remote than those who collect the data.

Finally, one may feel with a paper map or even with many models and statistical analyses that one can "get" the whole thing. But when it comes to a geographic information system, the "whole thing" is never revealed. It is always, in the end, there inside of a black box. And to trust the system, to believe its claims about the relationship of objects and elements to a whole, is to throw oneself blindly (granted, more or less blindly) into the fray. In the end, one is left—even with the most complex, multimedia-rich system—with no good reason to believe that the claims of the system, about what belongs where, who belongs where, are other than the detritus of a set of calculations. Yet in our everyday life, in our concerns with households and neighbors, and with the natural world, we clearly believe that something more is possible. Whatever else it can do, a geographic information system cannot provide that.

On nature

One might want to say here, "That's all very well for cases like geodemographics, but what about the vast number of geographic information systems that deal simply with the natural world?" Might one not admit that a geographic information system will have great difficulty in adequately representing neighborhoods and other places, but at the same time claim that the systems are perfectly fine in other realms?

Here it will be useful to turn to what is surely a commonplace in discussions about nature, and that is the idea of nature as a "web." This is of course only the latest of a series of images that have been used to characterize the natural world. For example, since the seventeenth century and the ascendance of modern physics it has been common to think of nature as a set of "dead" objects floating in space. And here it seems fair to say that the relationship between the objects and elements of nature are causal relations. But the idea of a web of nature refers back to other, older ideas, and especially that of the great chain of being.[36]

The idea of the great chain of being held that all of the elements of the natural world, from the highest to the lowest, most advanced to most primitive, formed an unbroken chain, with each connected to the next by a necessary link. To render an animal extinct would be to break the chain, to disrupt the entire order of nature.

If this idea seems quaint and outmoded, we in fact see it constantly in environmentalist literature, where it is claimed that to render the spotted owl or snail darter extinct, or to destroy the last vestiges of wilderness, would be to cause permanent damage to the natural world, to turn it into something else entirely. Indeed, it is in accounts of nature that we most commonly see the view that we are looking at a whole whose parts are intrinsically related to one another. One way, after all, of saying that two things necessarily go together is to say that "They just naturally fit," that they belong together.

Further, in the case of the natural, as in that of humanized places, the role of time is central. Preservationists and conservationists are painfully aware of the difficulties that arise in attempts to say when a landscape was "natural," and when it became artificial, or tainted. Here, Raymond Williams has written eloquently about the appeal of the idea that in some previous time nature was real, but that the humanization of the natural, over recent time, has effaced the imprint of the slower moving natural.[37] And so, in the case of that natural world we find raised, in the end, the very issues that were raised in the case of the social world.

So if it sometimes appears that as means of conceptualizing and representing the world geographic information systems are very general indeed, to the extent that they can be used to represent anything that can be geocoded, in fact the matter is more complex than that. In Chapter 1 we found that the systems—like all computer systems—seem to appeal to ideas of reason and language that were excessively rule-bound, and that lacked the ability to see either the situatedness of language and reason or the ways in which both operate in terms of human practices. In Chapter 2 we saw that the systems typically operate in terms of a complex of spatial conceptions, and that both a creator of the systems and a user of them will often engage in a process of switching, moving from one conception to another. And we saw that in doing so the user at the same time changes the implicit understanding of the relationship between the viewer and that which is being viewed.

Finally, in this chapter we have seen that beyond the process of switching there are several other processes at work in geographic information systems. Central to them all is the view that the world is in the end a world of information. We see this view elsewhere, in debates on intellectual property and in the way in which genetic material is spoken of as though it is simply a matter of bits, and not as some computer critics are wont to say, of atoms. But here the focus was rather different. If in Chapter 1 we saw that one can only understand language and reason by seeing their applications as matters of practice, here we saw that the very construction of places itself involves the establishment of sets of practices. And as a consequence, the reduction of the world to information doubly limits the ability of geographic information systems to represent the broad range of activities and elements that make up the world. In the end, this reduction recasts the objects and relationships in the world in contingent terms, and in a way in which the most basic of human relationships, like belonging, come to be derivative.

It is ironic that this is happening. In the 1970s GIS software could be written on a handful of punch cards, and the data could be an equally small set of eighty-column cards. Today both software and data have become almost unimaginably larger; a geographic information system today consists of what can only be described as a mountain of data. But as the amount of data has increased, the data have been transformed, from entries in a table or on a map produced by a line printer to remarkably rich representations of the world. Just as they have come increasingly to believe that the world is a world of information, those who create geographic information systems have endeavored to make their representations of the world more real and less like information.

Part II

GEOGRAPHIC INFORMATION SYSTEMS IN PRACTICE

4

ON THE ROOTS OF
GEOGRAPHIC INFORMATION
SYSTEMS

In an era in which one reads almost every day of advances in computer technology and of the application of computers to new areas of life, it may seem odd
to suggest that the history of computers is important. And we might say the very
same for geographic information systems; it seems, almost, as if they have moved
beyond history, beyond the constraints of their past. And yet if we are to understand the impacts of the systems today we need more than ever to understand that
history.

This is true for two reasons. The first is that in some respects the systems
themselves operate within contexts that are deeply constrained by history. This
will become most clear when we look at the issues of ownership and of privacy;
in both cases the ways in which those issues are understood have deep historical
roots, roots in part legal, in part political and in part cultural.

There is, though, a second reason for looking at the history of geographic
information systems. If we look at the history of the systems as it has been
written by those who developed and now use the systems we can get a window
into their understanding of what is important and what is not. We can begin to
see how *they* look at the social and ethical issues surrounding the systems.

On the conventional history of intellectual activity

There is a standard way of writing the history of academic disciplines, and this
way appears not merely in many full-scale histories, but also in the briefer
versions so often included in the introductions to textbooks. In geography we find
this in the "standard" disciplinary history, by James and Martin, just as we find it in
introductory texts.[1] In cartography we find it in Robinson, Morrison, Sale, and
Muehrcke, and in geographic information systems we find it, similarly, in
Burrough, just to name a single example.[2] Here the development of geographic
information systems (or of geography) is a matter of the move from an early stage
of primitive knowledge to a later stage in which knowledge is much more

accurate and highly developed. This development is one the logic of which is internal to the area of inquiry. This means that within a particular field, given an initial set of questions and theories, what we will tend to find is a move in the direction of the truth.

This way of writing the history of academic disciplines has much in common with traditional ways of writing about technology. There we typically find it assumed that if one starts with a particular tool or technological system, there will be an inevitable process of change, where the system becomes increasingly efficient. One way of characterizing this view is to see it as assuming that technologies are somehow "autonomous."[3] This approach is especially compelling when we are looking at technologies the accuracy or efficiency of which seems easy to measure, and the computer is surely such a system.[4]

Now, if we look at recent accounts of the history of geographic information systems it is not surprising that we find a combination of the two approaches. We find the history of a subdiscipline, and we find the history of a technology. In a moment I shall lay out the lineaments of that standard history, and I would hasten here to point out that it includes matter which is both interesting and important. Before doing so, though, I shall point to two features of such histories.[5] First, they by and large fail to meet the standards of "real" history, and largely because they tend so often uncritically to incorporate just such a teleological view of science and technology. And second, they nonetheless perform important functions for those involved in the areas of science and technology that are their subjects. This is because they are means by which disciplines create and perpetuate the myths and stories which help bind them together.[6] This is an important function, and not one to be dismissed; it does, though, suggest that in looking at these stories one ought to keep a careful eye out for the truly mythical, and more important, for the ways in which myths tend to exclude as well as include. Just as we know now to keep in mind the ways in which the founding myths of the United States, as laid out in textbook histories, have excluded important contributions by women and minorities, we should also be aware of the possibility that in the standard disciplinary history of geographic information systems much that is important remains unsaid.

On the history of geographic information systems

According to the conventional history, geographic information systems have been around for about thirty years. Although technologically complex, they are conceptually rather simple hybrids of two developments that occurred at that time. The first was the development of automated or computer-assisted cartography and the second was the development of models and statistical techniques for the analysis of data.

Like geography, cartography has, of course, long been associated with the state. As I noted earlier, at the time of the birth of Christ Strabo pointed to the close connection of geography with the state, and with its desires to use geographical knowledge in order to exercise more effective control over its territories; at the same time, surveyors and cartographers were consolidating that control through the creation of cadastral, or property ownership, maps, and through the inscription on the landscape of an abstract system of organization.[7]

In a very real sense, the history of cartography has been written as two parallel but separate stories. There is the development, from the early triumphs of Ptolemy in developing a picture of the earth, to a decline in the Middle Ages where the earth was represented in the starkest religious symbolism, to a renascence, with the rediscovery of Ptolemy, the age of exploration and then commerce, and finally the accurate mapping of the entire globe. And there is the second, and far more mundane, story, of the gradual development of increasingly accurate instruments for surveying, and the use of those instruments to move from the smallest line segments on the earth to larger and larger areas, gradually dealing with the roundness and then the lack of roundness of the earth.

The two stories started to converge early in the nineteenth century, as cartographers began to have increasing confidence that their maps of smaller areas were correctly located within the bigger picture,[8] but they do not really fully converge until the twentieth century. And indeed, even now it is not easy to move from a map of the globe through larger and larger scales and end up with a map of your backyard. If it is possible through satellite technology to determine what is cooking on the barbecue, it is much less easy to locate the barbecue on a nested set of maps, each covering a larger area than the last. This, though, is the very image of which cartographers have long dreamed, and it is the image which the development of automated cartography appeared to render possible.

Traditional cartography was very much a branch of graphic design. Creating a map involved the scribing of lines onto sheets of film by hand, the peeling away of parts of the film to create transparent areas, the manual location of previously typeset place names, and ultimately the sandwiching together of sometimes a dozen or more sheets of film in order to create negatives from which printing plates could then be made. Errors created at the wrong stage could sometimes mean that the entire process needed to be done again.

The earliest automated cartography attempted to routinize some of this process in a way that would at the very least make it faster and less prone to errors.[9] But it was not exactly a smooth process. David Rhind, for example, gives the following example,

[T]he name placement unit was fed with items from a gazetteer on punch cards, these having been previously selected on user-specified criteria using a mechanical sorter. The place name alphanumerics were flashed down a zoom lens in the correct font and any potential overlaps were obviated by an operator who was strapped into the device with a shoulder harness. . . . Since the great bulk of the operations had to be carried out in the dark and since the prototype machine seemed electrically unsafe (at least when examined in 1969), it was less than a total success.[10]

Even into the late 1960s, as this suggests, the primary goal of automated cartography was to be able rapidly and efficiently to create maps that looked like those created by hand; automated map editing tools were oriented toward the editing of individual map sheets. The cartographer's dream, of being able at will to derive individual sheets from some larger "virtual" sheet, was still just that, a dream.

In several senses another source of geographic information systems was the quantitative revolution within geography.[11] In the mid-1950s a group of geographers began to develop what they took to be a new view of the discipline, one in which geography was to be concerned less with regional differences than with the interactions of objects and events in space. The most extreme version of this view was called "social physics;" there an attempt was made to develop rigorous theories of interaction, modeled on the theories of physics. Within social physics the guiding image was of individuals and variables, arrayed on an isotropic plane, moving, agglomerating, and interacting according to discoverable laws.

The social physics movement is perhaps most important just because it provided to a generation of young geographers a new and emancipating image of what the field might be like. That generation, in some ways less visionary than the social physicists, proceeded to develop an arm of the discipline which for a time, perhaps from 1965 through to 1980, seemed firmly in control of the future of the discipline of geography, which they promised would now, finally, become a true science.

The quantitative revolution provided the second element for the invention of geographic information systems: A set of tools for the analysis of spatial data. But it provided something else; many of those involved in the quantitative revolution are today involved in the development and promotion of the systems.

In the early 1960s the two trends came together in several areas, first in Cambridge at the Harvard Laboratory for Computer Graphics and in Canada with the development of the Canada Geographic Information System (CGIS). And by the 1970s a large number of the systems had been developed, often in universities, but in close collaboration with state and national governments. Combining the elements of automated cartography and spatial analysis, the systems were typically conceptualized as means for dealing with *information*. They were systems for acquiring, sorting, storing, analyzing, and visually representing information.

In one sense geographic information systems were very much like any other databank. But by adding to their data a spatial component—a point, line, or area associated with the data—they were able to develop a much more powerful system. This is because in order for the standard databank to become truly universal, it would be necessary that each individual have its own unique identifier. A system that aimed to track the income of each person in the world would need to have, for each individual, a number unique to that individual. But by associating characteristics with spatial locations it became possible to develop a databank that allowed the generation of results that associated characteristics, by inference, with individuals who were not known. In a geographic information system it was then possible to create a system which looked universal even in the absence of universal data. And it is this visual element, this possibility of universality, which has been so central to the rapid development and dissemination of geographic information systems.

Beyond the conventional history

As a set of facts, this history is perfectly all right. It, or something like it, surely captures much of the flavor of a certain part of the development of geographic information systems, of the pleasure at seeing work that was once extraordinarily difficult, even impossible, become first only difficult, then easy, and finally routine. And yet, much that is important is missing here. Some might argue that this history focuses too much on the development of geographic information systems as problem-solving tools, and gives too little attention to other, more purely intellectual motives for their development. Others might complain that it pays too little attention to the ways in which the systems have provided means of entrée into employment.

But it seems to me that such objections, like the history itself, miss a central issue, and one especially important to an understanding of the social impacts and ethical issues surrounding the systems. Missing are two related elements. The first is an engagement with the relationships among academic science, government and business. It may seem odd that I say this; after all, Dangermond and Smith write explicitly about "The nature of the role played by a commercial organization,"[12] and the role of the Canada Geographic Information System is prominent. But it is one thing to attend to the existence of government and industry as *actors* within a system, and quite another to attend to the ways in which as actors they effected a reconfiguration of the *system itself*. Yet in the case of geographic information systems, this is precisely what happened.

This failure to consider the ways in which institutional changes have been generated is related to a second issue. This conventional history may seem, given its attention to personal details, to be one that attends not only to institutional factors but also to the personal; for that reason it may appear to be one which gives room to serendipity. Yet behind this surficial serendipity lies a view of the development of geographic information systems wherein that process is the fulfillment of a built-in essence. It appears there that the development was preordained to occur in the way that it did, with greater and greater computing power, greater and greater control of detail, and increasing standardization, all directed toward the ideal of one unified, if virtual, map. Here the development of the systems is seen as an autonomous process, one driven by this built-in image. And the writing of the conventional history in this way is implicitly a means of supporting this view, of justifying what has gone before and what is to follow.

In the remaining chapters of Part II, I shall begin to lay out the lineaments of a more thorough account of the development of geographic information systems. I shall show that in order to understand the place of geographic information systems in society we first need to see the ways in which the practice of geography, of using the systems, has changed. I shall show the ways in which the development of the systems has undercut traditional images of the relationship among those involved in scientific work. In doing so it has set the stage for a new conceptualization of the rights and obligations which attend being a scientist, and which, indeed, attend the use of a geographic information system.

When I say that I shall be laying out the lineaments of a more thorough account I mean just that. I do not pretend here to be synoptic, and indeed, would argue that it is a bit early in the game to be writing a history of geographic information systems. It is not, though, too early to descry the sweep of changes that are accompanying the development of the systems, and that is my aim.

5

THE RESHAPING OF
GEOGRAPHIC PRACTICE

Within what I have termed the "conventional history," there is a common way of thinking about science, and about research more generally. On that view the researcher is interested in the acquisition of knowledge, which may be theoretical or factual. The researcher comes to the world with a very general approach, called the scientific method, which involves the establishment of hypotheses, the acquisition of data, the testing of hypotheses, and so on.

From the usual perspective this practice of scientific research is in several ways different from other activities, and scientific expertise is different from other forms of expertise. The everyday activities of the scientist are different because of the particular set of values the work of the scientist can be seen as embodying.

Robert K. Merton, in the most famous and influential explication of this view, argued that scientists operate in terms of four such values. Those values are universalism, communism (or communalism), disinterestedness, and organized skepticism. According to Merton, scientists adopt these norms, perhaps knowingly, perhaps not, as they become socialized into science.

Universalism refers to the desire of scientists to attain knowledge that is universal. In the case of social scientists the aims may be more modest, and the desire may be merely to develop "meso-level" explanations, but in either case the goal is to move away from the particular. Because it appeals to a term like "universal," this goal puts science in the camp with other universalizing activities, like myth and religion. And like myth and religion, science embodies an overtly evaluative element, since there is the suggestion that the truths attained by scientists are in some way "better" than those that result from more particularistic enterprises, like history; at the same time, this ideal seems to suggest that scientists reside rather "higher" in a hierarchy where the quality of knowledge is the overriding concern.

Disinterestedness is not, Merton says, merely an attitude. Rather, it is a distinctive pattern of institutional control of a wide range of motives which characterizes the behavior of scientists. For once the institution enjoins disinterested activity, it is to the interest of scientists to conform on pain of sanctions and, insofar as the norm has been internalized, on pain of psychological conflict. . . .

Otherwise put—and doubtless the observation can be interpreted as lese majesty—the activities of scientists are subject to rigorous policing, to a degree perhaps unparalleled in any other field of activity. . . .

The translation of the norm of disinterestedness into practice is effectively supported by the ultimate accountability of scientists to their compeers. [1]

Skepticism, about which Merton says the least, is like disinterestedness both a cast of mind and an institutional structure. To be skeptical is to value—in a sense, above all others tools—reason as a tool, one that is universal in scope and that enables the scientist to further the evolution of science. It is to reserve judgment. And a skeptical institution is one that supports just that cast of mind, one that structurally supports a conservative stance toward the new.

Communism concerns the attitudes of scientists toward their work. The facts and findings of individual scientists are expected to be made publicly available, and to be generally knowable and manipulable. Because scientific works have an unusual characteristic—when one "uses" the works of others (at least, so it is claimed) one does not thereby "use them up"—it is possible that a scientist get credit for findings, while at the same time, "the substantive findings of science are . . . assigned to the community."[2]

Now one very important feature of Merton's analysis is the way in which it gives an account of the relationship between professionalism and expertise. Here, built into the social institution of science is a set of norms that force compliance on the part of the individual scientist; not to comply is to be deviant, and in the end no longer to be a scientist. And at the same time, there is implicitly a story about the way in which science developed as a form of expertise, and in which it found a home in society through the development of science as a profession. According to this view the institutions of science have functioned to create a kind of barrier or buffer between scientists and the demands of a capitalist society.

It has, of course, long been recognized that this view of science has a rhetorical and ideological function; in this it tends to be one with claims by other social groups to be above the fray.[3] Indeed, the very idea of a "profession" itself has a long and not-so-straightforward history. As Bruce Kimball has shown, where a "profession" was once simply a religious vow, the term came later to refer to the group that made the vow, then to non-religious groups, and only then to a set of "dignified occupations." Only recently did the reference of the term "professor" become limited, referring only to those in education, while the reference of "profession" itself became gradually broader, and in the twentieth century came to refer to a wide range of vocations.[4]

If some have pointed to the ideological and rhetorical functions of the term "profession," a number of scholars have also pointed out that in some cases— architecture is an example—the relationship between professionalism and expertise was reversed; the professions developed first, and only then did they attempt to define just what they were experts at.[5]

Still, it strikes me that as a set of defining images these "norms" do ring true. When scientists are asked about what they do these are the sorts of answers that they formulate. And it seems equally true that the introduction of changes that undercut the possibility of appealing to these images needs for that reason to be seen as involving a fundamental change in the possibility of justifying the scientific enterprise, and hence in the practice of science. This, in fact, is what has happened in geography as a result of the introduction of geographic information systems. We now find all of these values to be contested, in different ways. In this chapter I shall discuss the first three, universalism, disinterestedness, and skepticism. In the next chapter I shall turn to the last of these values, communism.

Universalism in the brand-name laboratory

If we consider the image of universalism, we see that Merton focused on theoret- ical results, on the way in which science typically aspires to the creation of accounts that are generally applicable. But there is a second way of appealing to universalism, and one that will be of more immediate interest here. When one engages in scientific work one is imagined to be moving away from the particular to the general not simply in one's theoretical pronouncements, but also in one's methods. So when I do an experiment it is important that that experiment be replicable, and it is also important that I be able to characterize the experiment itself in general terms. Indeed, if all I am able to say about an experiment is "I put the sample in the analyzer and pushed 'Process'," then I would hardly qualify to be called a scientist; rather I would more likely be termed a "technician," and unless there was someone who could characterize what I did in more general terms, the work itself would not qualify as science. Furthermore, one of the ways in which we test scientific results is by attempting to replicate our results using very different apparatus, and this, too, requires that we be able to specify just what each does, in general terms.

Now, there is a sense in which what I have just said is an exaggeration. After all, the scientist goes about every day using all kinds of tools, while knowing little about them. We use pencils without knowing from what trees they were made, or what the source of their graphite leads is.[6] We use pens without knowing what the chemical composition of their ink, let alone their barrels, is. And we use electrical and water and sewerage systems without really understanding the intricacies of their operation.

So in the case of the tools of scientific work ubiquity matters; at least some tools are so common that they are transparent, that their use is taken to require neither explanation nor justification. But when we move into the use of a second group of tools matters are more difficult. These are tools that are not at all ubiquitous, and further, that are quite particular, in the sense that they are created only by a single manufacturer, or a small group of manufacturers. We have of course seen just this for a number of years in chemistry and biology, where the complex equipment used for chemical analysis is manufactured by only a few companies.

Indeed, in writing up one's research results in these disciplines, one commonly mentions the manufacturer of the equipment used, just because this can be important information; some equipment is more and less reliable, more and less idiosyncratic. Moreover, with that information in hand a researcher can attempt to replicate results using different equipment, and thereby, it is generally believed, strengthen (or undercut) the results of prior research.[7]

We may in fact be tempted to imagine that this is just what is at issue in the use of geographic information systems. But in fact, the matter is different in several important respects. Most important is this: That in a given research setting the systems, and here I include the software and the hardware, take on a monopoly status. They are so complex and difficult to learn, and so expensive, that it is unusual for an individual routinely to use more than one.

If this distinguishes geographic information systems from other forms of scientific apparatus, it might, nonetheless, appear that in this respect the systems are more like, say, astronomical telescopes. Both, after all, are large, expensive, and complex means for doing research. Yet telescopes, too, are different in what turns out to be a vital respect. For the most part they are designed by laboratories and built to specification. They are "one-off" pieces of equipment. By contrast, geographic information systems are typically commodities, purchased from one of a few vendors.

And more important here—indeed, this is the fundamental point—the ways in which they are acquired differs fundamentally from the ways in which telescopes and other pieces of apparatus are acquired. Among suppliers of computer systems in general, and geographic information systems more particularly, there is a widespread practice of either donating equipment to universities or providing it at rates that are remarkably low, sometimes only 10 per cent of the manufacturers suggested list price. And so, for example, in an account of the establishment of a new GIS laboratory at the University of Denver we read "The generosity and support of the hardware and software vendors . . . must rank among the highlights."[8]

Those of us who have benefited from these arrangements sometimes find it difficult to appreciate the difficulties that attend them, but those difficulties are quite real. The initial difficulty arises because by providing products at a reduced cost or at no cost the producers are creating a constricted environment. One way in which it is constricted, and one that attends any major purchase of equipment, is that putting that equipment into operation requires a substantial investment of time, money, and space on the part of the organization receiving that equipment. Having made those investments an organization is extremely unlikely to want to give them up. In the case of computer systems, where the learning curves may be quite steep and where system operators may be expert only at a single operating or application system, there is an additional investment; it may be legally and morally difficult to release an employee, just because one has changed computer systems.

But against this background the initial and continued provision of equipment at reduced rates raises additional problems. An obvious one is this: people who learn to use a particular system typically feel most comfortable with that system. When they leave the university they are therefore likely, when asked, to recommend that that very system be purchased. And indeed, this is a primary reason for the existence of such corporate programs; if they are forms of benevolence and of tax savings, they are also forms of marketing. This may be all right, although when one is trained using a system which has been picked because it is inexpensive or free, and not because the trainer believes it to be the best, one needs to wonder.

But there are more difficult problems. Consider the following: Imagine that you are a researcher who discovers that International Computers and Environmental Systems (ICES) has been hiring illegal workers, exposing them to toxic wastes, and then as they prepare to complain, having them deported. Imagine, too, that ICES has made substantial donations of computers and GIS software, and that it is ready to make a further and substantial donation one week from now. Would you release your research results today? Or would you wait until the donation was secure? Most of us would wait.

This may not after all be a likely scenario. But the question that it raises is fundamental; when universities come to rely in substantial ways for support from industry (or, indeed, government), their autonomy is put into jeopardy. Indeed, in the above example lingering questions are sure to be raised. What if you chose to study companies that made video equipment instead of companies that made computer equipment? Might one not suggest that your very choice of research project had been affected by a desire, hidden or not, not to jeopardize the arrangements with ICES? And how might you counter that suggestion? I would suggest that you could not.

And so, notwithstanding the fact that many of the corporations that make these donations do so out of a genuine desire to support educational institutions, such donations introduce new and particular relations between universities and industry, and in doing so undercut the image of generality and impartiality that those in universities have used as a means of supporting their work. They inescapably leave the impression that generality has been replaced by particularity, impartiality with partiality.

And this is not the end of the matter. For I have allowed the assumption that there is some computer system, some geographic information system, that is indeed universal in its structure. But my discussions of reason, language and space have suggested one set of reasons for being skeptical of this possibility, and here I suggest a second set of reasons, more historical in nature.

As we have seen, there has long been a division of labor in geography, in information systems, and in cartography. Each of these areas has confronted projects large enough to require managers as well as workers. In fact, each has also had workers whose work was considered to be labor, remunerable in money, and workers whose work was considered to be the product of thought, even genius; work that required a different form of compensation, in the form of fame—or at least permanent credit. And so, the developer of the clock that allowed the accurate measurement of longitude received fame; those who carried timepieces for Mason and Dixon received only money. Mercator and Peters received eponymous credit for their projections; those who applied their projections in the production of maps got paid.

In order to understand the nature of changes in the division of labor, we need to return to what has in science been a single most powerful image, one that has functioned most to guide discourse about science, and also about cartography. This has been the image of the thinking head and the laboring body. This image has been of extraordinary importance, just because it is right at the center of discourse about the place of the computer and now the geographic information system both in the system of labor and in the system of property.

Of course, not all thinking *counts* as thinking. The temporary immigrant living in a dormitory and writing lines of computer code is surely engaged in thinking, yet this work, and the work of the digitizer, is viewed as a species of labor, and grouped with that of the tool and die operator.[9] The difference is this: This sort of intellectual work is seen as routine, habitual, interchangeable—and for that reason, so too is the person who does it. It isn't stretching the matter to say that some people see it as almost animal in nature.[10] So it is not merely thought that counts; it is creative thought, thought that is somehow beyond routine. Or at least, this is the traditional view, and one that has come to be seen as common sense since the decline of craft work and the rise of the factory system.

But the development of the computer has thrown a wrench into this way of looking at the matter of rights and responsibilities. The image of the head and the body has been built into so much of the discourse about computers that it has come to seem quite natural. Indeed, we see it when we first face the screen. By and large—although as we shall see this is changing—we begin by issuing a "command." We then speak of the computer as having "executed" the command. So the person operating the machine has used "thought" and "intention," while the machine itself has merely "obeyed" the order. Yet to say that a machine obeyed a command is to anthropomorphize. It is, after all, to suggest that the machine received the command and made a decision to act upon it.

Now, recent work on the nature of natural languages tends to agree that people learn language by learning practices; they learn the rules only later. Moreover, the "rules" of a language are never sufficient to capture what appear to its everyday users to be a language's manifest regularities. And if specialists cannot write the rules by which we speak, why should *we* be expected to be able to do so.

But here the case of computer languages may seem sharply different: In the case of FORTRAN or BASIC or LISP the rules *are* written down, they are explicit. These languages are invented, from the ground up. When I say "GO" in FORTRAN I mean "GO," and nothing else; when I say "Print" in BASIC I mean "Print," and that is what the computer will do. This argument strikes me as a compelling one. Leaving aside mistakes, which at least in a very simple language could almost certainly be excised, these languages *are* regular, and *do* involve the application of sets of rules.

Yet there *are* schools of programming practice, just as programs may achieve the same end via means alternatively described as elegant, sloppy, or efficient. One can be a poet or a hack in FORTRAN as in French. Still, the simplified image of language is persistent, not merely among scientists but also among those involved in controlling the production of computer software. Turkle and Papert, for example, have shown this in a discussion of the difficulties involved in developing and promoting alternatives to traditional styles of programming.[11]

Similarly, in Judy Wajcman's related analysis of the masculinization of computer culture, we see—in a way closely related to that propounded by Wittgenstein, and then later by Langdon Winner—that the development of styles of thinking that emphasize the ability to formulate explicit rules is supported not merely by the silicon and steel which make up the machinery.[12] It is supported, too, by the image of the machinery. And more important, it is supported by the articulation of the computer into a pre-existing edifice of production.[13]

And in his *Forces of Production* David Noble has documented the ways in which industrial automation has chosen a route that supports the authority of white-collar technicians' explicit knowledge, where it might have chosen a route that supported blue-collar machinists' practical knowledge.[14] Finally, David Sudnow has shown in the case of video games that success requires a kind of skill that could never be learned from a book, or indeed from the application of any set of rules, but only through the acquisition of sets of skills.[15]

What we see in each of these cases is the articulation—by the department managers, plant owners, and parents of which these authors write—of a distinction between patterns of activity that are seen as being what I would term "practice saturated" and those wherein it is believed to be possible in principle to codify those practices into formalized sets of rules. Turkle and Papert show that managers saw hierarchical styles of programming as more orderly, and more intuitive styles as being too messy, too unpredictable to be allowed in a workplace. They suggest that using object-oriented programming, for example, one can take a different approach, and one that sees the act of writing programs as a matter not of the application of rules but rather as the use of certain skills and intuitive understandings. But, they conclude, it is those who employ the programmers who decree that in order to be scientific and business-like—to fit the appropriate image—those who write programs must do so in a hierarchical fashion and must appeal to explicit rules. It is this, Turkle and Papert conclude, which has tended to force the work of programming into a mold which in turn excludes women and those with alternative intellectual styles.

Wajcman shows that in our society cognitive styles associated with women are typically seen as practice saturated and are contrasted with those more "orderly" and rational styles associated with men. The gendering of the computer industry arises, she concludes, not from some essential mental differences between men and women but rather because men and women are so often and in so many ways taught to differentiate themselves sexually in terms of different practices.

Noble makes a similar argument, showing that an image of the "orderly workplace" was at the core of the adoption of digital, white-collar-controlled systems of numerical control in preference to available analog systems. The analog systems, in fact, began by presuming that the knowledge to be used was practice saturated, and attempted to take advantage of that fact. Those in charge of adopting the systems rejected them, and in doing so appealed explicitly to the non-practice-saturated images of digital control; indeed, Noble argues, the images so powerfully connoted control that the digital systems were adopted even though they were less efficient than the ones rejected. The appeal to explicit rules, then, rests on the long-standing appeal to explicitness, as in part an element of cost accounting and corporate control, which we now refer to as Fordism and Taylorism.

Finally, Sudnow shows that in the case of video games, as in all practice-saturated activities, real mastery occurs when one can "let oneself go," exploiting the knowledge that has been incorporated in the body. Indeed, in complex and time-constrained activities like playing video games—or flying an airplane or giving a news conference—those who need to "stop and think" about what to do next simply cannot succeed.[16]

There is, though, a double irony here, and the nature of that irony ought by now to be clear to the reader. The irony arises in the first instance because Foucault has shown that it is just this saturatedness that renders individual life so much under the control of outside forces.[17] An organization in which control was exercised only through explicitly named channels would be an organization close to anarchy, very much as a computer program is always close to the anarchy that can arise when a single letter is missing.

And in the second instance, it arises because, as should now be clear, every aspect of life is practice saturated. The image of a human activity beyond practice is simply an image, and in being so is the means by which the consequences of actions are denied.

In a wonderfully provocative work entitled "Work for the hairdressers: The production of de Prony's logarithmic and trigonometric tables," Ivor Grattan-Guinness makes this more clear, as he describes the development under Napoleon of the *Tables du cadastre*.[18] This set of tables, described as "by far the largest table-making project the world had ever known" (p. 11) took its structure, according to Grattan-Guinness, from Adam Smith's *Wealth of Nations*. There Smith had advocated an economic system based upon the division of labor; using the famous example of the pin-making factory, he had argued that the various tasks involved in making pins could best be carried out if they were distributed among different individuals.

> To take an example, therefore, from a very trifling manufacture; but one in which the division of labour has been very often taken notice of, the trade of the pin-maker; a workman not educated to this business (which the division of labour has rendered a distinct trade), nor acquainted with the use of the machinery employed in it (to the invention of which the same division of labour has probably given occasion), could scarce, perhaps, with his utmost industry, make one pin in a day, and certainly could not make twenty. But in the way in which this business is now carried on, not only the whole work is a peculiar trade, but it is divided into a number of branches, of which the greater part are likewise peculiar trades. One man draws out the wire, another straights it, a third cuts it, a fourth points it, a fifth grinds it at the top for receiving the head. . . . [19]

Applying Smith's principles to the construction of the new cadastral tables, de Prony divided the work into three parts. The first group consisted of a series of mathematicians, who determined which mathematical formulae were to be used. A second group, of managers, organized the computational work and compiled the results. And a third group, of "human computers," did the actual computations, which were simple enough to be done using only addition and subtraction.

This idea—of the division of labor—was in the early twentieth century formalized in works by Frederick W. Taylor, Frank and Lillian Gilbreth, and others.[20] There, of course, the idea of a division of labor was extended; their works established ways for breaking down complex patterns of human movement into simpler and simpler units, with the idea that the work process could then be reconstructed, more efficiently, from these "simples" (Figure 5.1).

Now, as Campbell-Kelly and Aspray, and more explicitly Phil Agre, have shown, a rereading of the history of office technology, through the development of the computer, shows that this model of work was right at the root of the creation of the first models for machine computation.[21] That is, the computer was not in the first instance modeled after an image of human thought, but of human labor and human practices. And it was human labor under a very specific set of economic conditions.[22]

This suggests, then, that the assumption that one ought somehow to imagine that one can readily and without explicit justification map the sort of processes that take place in a computer onto those that take place in humans—and hence in science—is simply erroneous. And, indeed, work on the history and sociology of computer programming leads one to the same conclusion. For example, Joan Greenbaum begins her "The head and the heart: Using gender analysis to study the social construction of computer systems,"

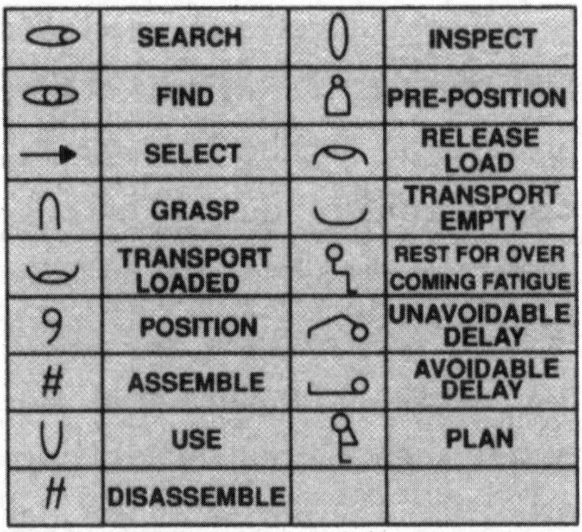

⊂⊃	SEARCH	0	INSPECT
⊂⊃	FIND	8	PRE-POSITION
→	SELECT	⌒	RELEASE LOAD
∩	GRASP	⌣	TRANSPORT EMPTY
⌣	TRANSPORT LOADED	9	REST FOR OVER COMING FATIGUE
9	POSITION	⌒o	UNAVOIDABLE DELAY
#	ASSEMBLE	∟o	AVOIDABLE DELAY
U	USE	8	PLAN
#	DISASSEMBLE		

Figure 5.1 Therbligs

Source: Michael O'Malley (1990: 236).

74

For ten years I worked as a programmer and systems developer. . . .
Then, for the next ten years I studied the work of systems development,
using new glasses to reflect back on the division of labor in the art and
science of developing computer systems.

But the division of labor I had experienced and the division of labor I
studied were indeed on different dimensions or planes. Using the tradition of Marx and Braverman, I examined the organization of labor as the
division between the 'head and the hand'—or the separation of tasks of
conception from those of execution. What I had experienced, was yet,
another dimension of variation—that between the head and the heart.[23]

For Greenbaum, the understanding of computer systems in terms of this
image, of the head and the heart, attempts to "show how the system development
process builds on the base of the natural sciences, and in so doing, borrows from
the gender-based myths inherent in the sciences" (1990: 10). And for her,
following Keller,[24] this in part involves an understanding of the linguistic
dichotomies that are so firmly entrenched in scientific thought, where objectivity,
reason, rationality, and power are seen as masculine; subjectivity, feeling,
emotions, and love are associated with the feminine. The conceptualization of the
labor process involved in the development and use of information systems tends
to be directed toward the first list, toward the masculine, in a way that reduces
the labor to its products, and in so doing renders invisible the processes by which
it is made. Putting the matter as I have above, this is a matter of conceptualizing
the labor process in terms of a set of rules which it is assumed must be involved
in the creation of those products. Here, "the *written word* is golden" (1990: 14),
and tacit knowledge and style of work are ignored.

Both Greenbaum and Wajcman implicitly argue that one cannot distinguish
clearly between conception and execution, and that it is surely an error to see
execution as following inexorably from conception.[25] In any given case there are
always a large set of rules that might apply; just as an unlimited set of equations
can connect a finite set of points, so too can an unlimited set of rules be made to
accommodate themselves to a set of circumstances. Indeed, as Alasdair MacIntyre
has argued, it is this very fact, itself an artifact of the same modernist ways of
thinking that gave us science, that has made life today seem so difficult, so full of
irresolvable moral dilemmas.[26]

So in different ways a wide range of authors suggest that the models of the
computer, of computer reason, and of scientific reason are closely, even intrinsically tied to capitalism, to industrialism, and to masculine culture. In their
arguments all suggest that the sort of universalism that Merton saw in science
cannot be assumed to carry over to the practice of science when that practice is
strongly imbricated with the use of computers.

On being disinterested in the porous ivory tower

If you look through an introductory textbook on manual cartography, something will no doubt strike you; at least it will strike you after it is pointed out to you. The practice of manual cartography requires the use of a large amount of equipment, from pens and compasses to scribers, to typesetters, then on to darkroom equipment, cameras and platemakers, and finally to printing presses. But from a textbook you would be unable to create a shopping list. True, it's often the case that there is a photograph of a piece of equipment, with credit given to the company that manufactures it. But nonetheless, you will find yourself rather at a loss if you try to learn the subject from a text.

At the same time, if you do take a course in manual cartography, or better, if you take several from several people, you will likely find yourself given information about a variety of tools. In one class you will use one sort of pen, in another class another. Instructors will have preferences for almost every type of product. Yet as you move from class to class you will find three things. First, there will be both differences and similarities in what is used. Second, you will be able to ultimately to develop a suite of equipment which you prefer. Third, though, and most important, your own individual choices will not make an essential difference to the products that you create. Switching from pens by Mars-Staedler to pens by Koh-I-Noor will not make an appreciable difference in your output.

In the matter of geographic information systems, though, things are rather different. First, it is possible to learn to use a geographic information system from a book that not only mentions particular products, but that is actually produced by a manufacturer, ESRI.[27] Here the GIS curriculum is specifically tailored to dovetail with the abilities of the product. And second, the tools that you use matter a great deal, and a commitment to one entails a commitment to another. Purchasing a certain computer limits your software choices.

More fundamentally, and more importantly, the software that you use limits the sorts of analyses that you can do. Now, this has one meaning when you are using a whole set of tools. But when you are working strictly with a single system it means something else again, as the focus inescapably turns to finding data that can be analyzed with the system and then performing the analysis. This, of course, is just what we find in the courses taught by vendors themselves, and here matters become even more complex when we note that at least one vendor certifies instructors, who offer courses within universities.[28] One comes easily and naturally to define one's research questions in terms of the capabilities of the system. In fact, here, as we shall find in the next section as we consider the establishment of systems of standards, the system in a very real sense comes to define what exists and what does not.

Now, it might be replied, and I believe quite correctly, that this is a feature of all models and theories, that by definition a model or a theory defines what exists and what does not. But with geographic information systems there is a difference, and the difference becomes increasingly pronounced as those systems come increasingly to arrive in the form of packaged products, of turn-key systems, and decreasingly to be constructed of algorithms and models and code supplied by the person using them. For the costs of changing increase dramatically as one knows less and less about the inner workings of the system. It is one thing to change a piece of an equation, another to rewrite lines of code. And it was one thing to rewrite that code when an entire GIS program was written on a few hundred punch cards, and when one already knew the language in which it was written. It is quite another when the program is really a system, of dozens of megabytes of code. There the costs of changing systems, in order to be able to frame a question in a different way, can be immense.

There is a final sense in which the increasing porosity of the university is important. It concerns what happens after the classroom. A quick look at advertisements for employment, at advertisements placed by those looking for employment in industry and government, and at curricula vitae of specialists in geographic information systems applying for academic jobs will show a striking thing; in each case there is explicit mention not only of desired skills and abilities, but also of the brand names of products. To take a single example, and one can find a great many simply by doing a search on the Internet, the California State University-San Marcos posted the following notice:

> Interdisciplinary: GIS, Geography, Planning, Public Policy. Liberal Studies Program at California State University-San Marcos seeks tenure track interdisciplinary Assistant Professor to begin August 1998. Teaching and research must include use of GIS (ArcInfo and/or ArcView). Ph.D. required. Areas of specialization open, but social and ecological impacts of development, relationships between unequal regions, or international borders preferred. Fields may include city and regional planning; environmental, resource, urban, industrial, or regional geography; and public policy.

Employers want employees who are able to use certain products. It is no surprise that prospective employees advertise those skills, and more, that they wish to learn those skills within the university classroom.

So what we see in the case of geographic information systems is the establishment of a cycle, from procurement of hardware and software by donation and discount, through teaching in terms of proprietary products, through employment based on the knowledge of those products. The different parts of the cycle support one another, and support the development of curricula, departments, and universities closely tied to the products of one or a small number of vendors. Here the various actors develop symbiotic relationships one with the other.

We find a similar pattern if we turn from teaching to research. Not only do they donate hardware and software, some companies also make donations to universities to support research. Now, both government and foundation support for research have long histories, and the perils of each are well known. But when a corporation is funding research to be carried out using its own products an additional concern arises. The situation is, after all, rather like that when one uses a vanity press, paying to have one's work published. There always remains the underlying feeling that the work was not good enough to make it on its own.

Here, too, the issue of censorship arises once again. If my work is funded by ICES will I write results, or even pick a topic, likely to embarrass them? In the case of work funded by ICE and using ICE's software, who holds the rights to the results, and who is responsible if they are somehow improper? No one would claim that if I made an inaccurate map using a Koh-I-Noor pen the manufacturer of the pen should be held accountable. Some might say that ICE should, if they have provided hardware and software that they claimed to be capable adequately to do a particular project, but which had hidden flaws. But what if the researcher was in effect paid by ICE to use the software, if it was an implicit condition of the grant?

Here, as in the previous cases, we find that the introduction of geographic information systems involves a recasting of the relationships between the university and industry. Individually none of this is new. After all, in the natural sciences many of the same practices have been common for many years. But in this case what makes the matter different is the rapidity with which computer and now geographic information systems have been introduced into universities, and the scale at which they have been introduced. Here we do not see a gas chromatograph being introduced into an existing laboratory; rather, we see the manual cartography laboratory being replaced, brick by brick, in its entirety.

As a result, in each of these cases, of the procurement of equipment, of the use of proprietary equipment in teaching, of the use of that equipment in research, and of the use of industry-funded grants in research, we see a refiguration of the relationships between industry and the university. For Merton this refiguration would be seen as a matter of the values of industry moving into the university and replacing the traditional values held there.

Here again, though, I would suggest rather a different way of putting the matter. What we see is a recasting of the images to which academics have appealed. Where they appealed to an image of universality, of the universal theory and the universal tool, we now see particularity everywhere, in the computers and software, even in textbooks and model curricula. We see grants provided by corporations where the research is carried out using the tools which those corporations make and sell.

None of this implies that suddenly geographers and cartographers have become corrupt. Indeed, I would argue that it would be a serious mistake to see this as the problem. Rather, what we are seeing is the development of a new structure of geographic practice, and one where the images which guarantee the quality of one's work are not images of "universality" and the like, but rather images of particular products, known to work.

Here the discourse in which one engages leads one in very different directions. How this is the case will perhaps best be made clear through an analogy. Imagine yourself to be an average person who knows very little about automobiles, other than how to start them, put gas in them, and take them to be repaired. As such a person you are likely to have very little to say about automobiles, other than about how they look and whether they are working. But if you spend some time learning about the operations of the various systems of which they are composed—the starting system, ignition system, braking system, engine, transmission, steering system, and so on—you will find several things happen. One is that suddenly a whole range of discourses that once seemed like gibberish now make sense. You can now understand automobile magazines when you are stuck with them in the dentist's office. Further, you now can understand something about a whole range of social groups which were once quite foreign. You feel more comfortable in a service station, and understand something of what you read in the sports section of the newspaper. Learning about automobiles involves learning a new way to talk, and in doing so implicates you in a range of groups of which you were not a part.

Much the same has happened with the introduction of computers and geographic information systems into universities. Now, suddenly, one can find oneself making sense of a wide range of discussion, by people in the computer industry, in telecommunications, in engineering, and on and on. One is thrust outside the university, just as the computers have thrust themselves in. And this is expressed in all manner of ways. Inexpensive academic conferences give way to expensive "business-priced" conferences. Important research is only available in the fugitive proceedings of those conferences. Book-length studies are published not by university presses but by business and technical publishers; they are sometimes full of advertisements.[29] The barrier between the university and the "outside," a barrier about which advocates of universities and of professions more generally have made so much, is now porous.[30] And as it has become so, the structures that have supported the value of disinterestedness have come apart.

Skepticism and democratic ontology

Merton's third value, of skepticism, is also being challenged. In order better to understand the issues here, it will be helpful to look to a related matter, to the development of the printing press. We all know that the printing press was developed in Europe in the latter part of the fifteenth century. And it is a commonplace that it was a watershed development. In a fundamental way it allowed for the development of literacy, because for the first time books could be widely disseminated and, indeed, the newspaper would be quite impossible without it.

But the printing press allowed something else. Before the press, each copy of a document introduced new errors, and to check those errors required that one quite literally go to the original. After the printing press one was able to compare a new version of a work with a standard printed one.[31] For the first time it was possible to have documents that were identical; indeed, it now became possible for the first time to imagine the creation of an edition that was "definitive."[32]

We are now so used to what this means that it is hard to see it. But the notion of two texts being "alike" involved matters as basic as the agreement that two squiggles on a page, such as 'a' and 'A', are the same latter. More fundamentally, it involves a theory of reading, that sees the letters 'cat' and 'CAT' as leaving the same impression on the reader's mind. When we read a book in which the text is formatted in "Bujardet Freres" rather than "Garamond" or "Bookman" or some more traditional typeface, we still—on this view—have the same thoughts, we still read "the same thing" (although we may express irritation with the person who chose the typeface).

Similarly, although here we begin to deviate from the concerns of the person intent on producing a definitive edition, in our everyday reading we gloss over differences in spelling, treating "standardize" and "standardise" as though they are the same word. And we generally do the same with typographical errors, imagining that we know what the author meant. Here too we have a notion that the written squiggles on the page, the letters and words and sentences, have a direct meaning to us, a meaning that doesn't require some intermediate act of translation or interpretation.

Now the point here is that the very possibility of a notion of "standard English" rests on the development of a particular technology, of the printing press. The press allowed the rapid dissemination of the same material to a wide range of people, at first within limited geographical areas, later within much wider areas. These, as Benedict Anderson has shown, gradually became "national print languages" and without the printing press such a concept would not have been possible.[33]

Now, if we turn back to maps we find similarities and differences. Certainly maps have long been mass-produced. According to Arthur Robinson,

> The first printing, five hundred years ago, in 1472, of a simple allegorical map was an important event in the history of communication. It ranks a close second to that of printing from movable type, a develop-

ment that occurred only some twenty years earlier. To appreciate properly this momentous event, we must remember that in all preceding time maps had existed only in manuscript form.[34]

Robinson continues

The capability of printing maps immediately opened the way for countless numbers of exact duplicates that, for the first time, allowed scholars easily to compare many geographical portrayals, consider their characteristics, and plan ways to produce even better images of the emerging world.[35]

But note here what Robinson does *not* mention; this did *not* lead to the development of a "standard cartographic language." Rather, all that could be hoped for were "better images of the world."

Even granting that, as the eighteenth and nineteenth centuries saw nations embarking upon large cartographic projects and saw cartographic establishments as a consequence developing sets of standard practices, something was still missing. What was missing was a way of combining existing maps, of redrawing them at different scales, that did not pose the very same threat that faced the medieval writer of manuscripts. In every copy there was the danger that something wanted was left out, something unwanted added.

This was true in one way when changing the scale of a map involved the use of mechanical devices, like calipers and pantographs. And it remained true after the development of optical apparatuses for projecting larger or smaller images of maps. But even after the development of photographic means for enlargement and reduction there remained problems. Line widths, symbols, and especially typography do not easily enlarge and reduce; moreover, just because of the nature of projections, sheets seldom match.

It nonetheless has remained that cartographers have believed that the history of their work has been one which has involved the gradual reduction of error, the ever increasing ability to create better and more accurate maps of the world, where the elements in the map refer more and more directly to the entities in the world.

Yet, perhaps ironically, the development of geographic information systems seemed to undercut that belief. For when one attempted to combine different systems all kinds of difficulties arose. One set of difficulties were predictable, and not necessarily related to this issue of accuracy. Different systems used different methods of coding and storage. Just as one has difficulties running a program written for a Macintosh on a machine which runs under MS-DOS, in an era when most systems were one-off combinations of hardware and software, trading data was difficult indeed.

But there were two other sets of difficulties, which spoke more directly to the issue of quality, the problem of quality itself, and the problem of features and objects. As it turns out, the solution of the problem of the interchange of data required the parallel solution of the other two problems. The solution to these three problems has been the establishment of a series of standards, and it is these standards which seem to hold the promise of doing for cartography what the development of the printing press did for written language.

Here, though, we shall see once again that these developments involve a refiguration of the relationships among science, industry, and government. Moreover, we shall see that they do so in a way that shares some of the more problematic features of the development of standards in language.

In contemporary life "standards" are both ubiquitous and invisible. Indeed, most of us would be quite shocked to see how thoroughly our lives intersect with standards of various kinds. Moreover, for those whose first reaction to their existence is to see it as, once again, an example of the ever-growing reach of government, there is another shock; a great many of the standards which guide contemporary life are established by private organizations.[36]

In the case of digital cartography and geographic information systems, there are actually several interlocking (and sometimes competing) standards. Especially important have been the Federal Information Processing Standard (FIPS) 173, the International Organization for Standardization (ISO) 9000, and the ongoing work of the Federal Geographic Data Committee (FGDC), now associated with the National Spatial Data Infrastructure (NSDI). Also involved, though, have been the American National Standards Institute (ANSI), the International Electrotechnical Commission (IEC), the European Committee for Standardization (CEN), the European Committee for Electrotechnical Standardization (CENELEC), the British Standards Institution (BSI), the United States National Institute for Standards and Technology (NIST), and the American Society for Testing and Materials (ASTM), and many others.

Perhaps most important in the United States has been the development of FIPS 173. In fact, it has developed a conventional history, one which has been recited in more or less the same form by a number of writers.[37] It began with the creation in 1982 of the "National Committee for Digital Cartographic Data Standards" (NCDCDS), under the auspices of the American Congress on Surveying and Mapping (ACSM), but by request and under funding from the Department of Interior's United States Geological Survey (USGS). Closely connected was the United States Bureau of Standards (now the National Institute for Standards and Technology (NIST)).

A year later a second organization joined the fray; it was the Federal Interagency Coordinating Committee on Digital Cartographic Standards (FICCDCS), formed by the United States Department of Management and

Budget for rather a different set of reasons. Where the NCDCDS had been directly interested in developing cartographic standards, the FICCDCS was charged with eliminating inefficiency and duplication, and the establishment of a set of standards for geographic data was merely a means to that end. Ultimately the Standards Working Group (SWG) of the FICCDCS developed what it called the Federal Geographic Exchange Format (FGEF).

The work of the two committees was merged, through the work of the Digital Cartographic Data Standards Task Force (DCDSTF), in a 1983 "Proposed Standard for Geographic Data." This document was published in draft form by the ACSM.[38] The largest part of the document consisted of the Spatial Data Transfer Specification, part of which, related to implementation, consisted of ISO 8211, which also has an ANSI number (also 8211) and a FIPS number, 123.[39]

The proposed document was set for testing by the USGS, and in final form, now called the "Spatial Data Transfer Standard," it was approved 29 July 1992, and renamed FIPS 173. At that point the USGS was named the "maintenance authority" and it moved toward implementation. It became effective 15 February 1993, and on 15 February 1994 its use by Federal agencies became mandatory. This is not, though, the end of the story; the actual application of the standard, itself rather an abstract and general document, requires an initial specification. Those planning to work in a particular area develop individual "profiles," which make explicit the ways in which it will be applied.

From its name, the Spatial Data Transfer Standard, one might expect just that, a system for making digital geographic data which exist in one format available in another. It might, that is, appear merely to be a means for eliminating the waste and duplication which arise when different companies or agencies create maps or databases which cannot be translated one into the other. This, in fact, was precisely what had happened during the early development of geographic information systems, where each system was unique, and where, frequently, there was little in the way of documentation to guide the person interested in combining systems.

But in fact, the Spatial Data Transfer Standard involved much more than that, for the transfer of data implicitly meant not merely that material stored on one type of machine be readable on another; it also required that the machines be talking about the same thing. In effect, the standard required that one create a standard language, which included an ontology, a theory of what exists. At the same time, it required means of determining not only whether a river here was a river there, but also whether the person (or machine) here had exercised the same standard of care that the person or machine there had used.

The introduction of the issue of quality, though, implicitly (as in the case of FIPS 123) brought into play an additional series of actors. This is because any national standard, and especially if it is going to be adopted on the international market, needs to comply with the ISO standards. These standards have come to be *the* standards of the European Union, and their reach has now extended to a total of ninety member nations (including the United States, in the form of the ANSI), and thereby, for all practical purposes, to the entire world.

The ISO standards are produced by the International Organization of Standardization, a private organization established in 1946. It produces a wide range of standards, some more and some less official. Among the most well-known has become a family of standards termed ISO 9000, all broadly concerned with the issue of quality. ISO 9000, published in 1987, was created by a committee (Technical Committee 176 on Quality Management and Quality Assurance) formed in 1979; the United States was represented on the committee by the ANSI. ISO 9000, in turn, was based upon a British standard, BS 5750, which itself was in turn based on a British military standard, MIL-Q-9858 (which had been adopted by NATO as AQAP-1).[40]

ISO 9000 has been formally adopted by the European Union, and by a total of fifty-six countries. Now, this does not necessarily mean that each manufacturer in each industry covered by ISO 9000 needs to comply with it. In some cases compliance is virtually mandatory; within the European Union certain products are required to meet safety standards, and compliance with ISO 9000 is considered to be evidence that those standards have been met. Certainly any manufacturer who wished to export its products to the union would find it in its best interests that those products comply with ISO standards. In other cases, and certainly in the case of digital data and computer software, the adoption of the standards by government (as in the case of the adoption in the United States of ISO 8211, via FIPS 173) makes it appear to be in the best interests of manufacturers to create products that meet the standards.

This has been a long and complex story, the end of which, actually, I have left out. The development and application of these standards continues today through the work of the Federal Geographic Data Committee (FGDC), which was authorized in 1990 by the Department of Commerce. The FGDC has, in turn, promoted the development of a united view of geographic data, the beginnings of which were established by the 11 April 1994 Executive Order "Coordinating Geographic Data Acquisition and Access: The National Spatial Data Infrastructure." Among its goals was

> under the auspices of the FGDC, and within 9 months of the date of this order [to] develop, to the extent permitted by law, strategies for maximizing cooperative participatory efforts with State, local, and tribal governments, the private sector, and other nonfederal organizations to share costs and improve efficiencies of acquiring geospatial data.[41]

This, as expressed in the FGDC's "The 1994 Plan for the National Spatial Data Infrastructure: Building the Foundation of an Information Based Society," involved the articulation of further sets of standards.

This last part of the story, like the earlier parts, involves an alliance among academic scientists, government agencies, and industry. As we have seen, though, particularly important have been non-governmental agencies, like the ISO and the ASTM, and professional organizations, such as the ACSM and, particularly in the case of the NSDI, the Association of American Geographers (AAG).

Now, the interesting issue here is the place of scientists in the process. The NCDCS and DCDSTF each had steering committees, working groups, and advisory groups on which served a number of academics, as shown in Table 5.1. Were the role of the two organizations above—and the ratios of academic scientists to non-academics seems typical—simply the establishment of standards for the interchange of data, one might find here nothing on which comment was needed. But in fact, if we look at the Spatial Data Transfer Standard we find much that is more directly related to the other issues, of language and attendant ontology, and of quality. For example, in "Informative Annex A: The SDTS Model of Spatial Data," we find the following:

> The basis of SDTS is a model of spatial data sufficiently general that it can be accepted, but at the same time sufficiently structured to provide an adequate basis for the organization of spatial phenomena. . . .
>
> The SDTS conceptual model has three parts: a model of spatial phenomena; a model of the spatial objects used to represent phenomena; and a model of spatial features, which explains how spatial objects and spatial phenomena are related.[42]

More specifically, the SDTS elaborates a vocabulary of entities, that is, of things which exist in the world; a vocabulary of objects, or representational elements; and a vocabulary of relationships among the two.

With respect to the entities which exist in the world, they can be seen in two ways. A particular thing, the Brooklyn Bridge or Washington Monument, is an "entity instance." But more generally, "bridges" constitute "entity types." Both entities and entity types, in turn, have characteristics, like color, but in addition have instantiations of those characteristics, like "blue."

The development of this sort of vocabulary and of an account of the relationships among entities, objects and so on, is in the most fundamental sense a part of science; battles have been and continue to be fought over just the issues raised here. The intensity of these battles provides ample testimony that these issues are not without consequence. More to the point, scientists (and philosophers before them) have argued not merely that they have the answers to these questions, but

Table 5.1 Steering committees

Committee	Academic Scientists	Total Number of Members
NCDCDS Steering Committee	5	13
NCDCDS Working Group on Data Organization	3	10
NCDCDS Working Group on Data Quality	3	11
NCDCDS Working Group on Cartographic Features	2	10
CDCSTF Voting Members	0	8
CDCSTF Advisory Members	4	7
Total	17	59

that the asking and answering of these questions is fundamentally in their domain. Yet here, in the case of digital cartographic standards, the answers were reached in committees. Out of fifty-nine members, seventeen were academic scientists; the remainder were representatives of government and industry. We have the establishment of a standard which describes what exists and how. It is a standard promulgated in part by people whose livelihoods rest on their being able to market their products, products which come with their own built in theories of what exists. It is promulgated in part by government bureaucrats. And it is promulgated in part by members of professional organizations, each of whom has an interest in seeing that what exists be something that can be described using his or her tools of the trade. Here, perhaps more than in the previous cases, the boundaries that have in the past separated science, industry, and government have been not just breached, but eliminated. Here the institutions that supported skepticism are replaced with ones wherein what exists is determined by a vote.

6

WHO OWNS GEOGRAPHIC
INFORMATION?

The final value that Merton discussed, the value of communalism, seems on the face of it to be very much what is at issue in science, and perhaps the more so in geography, at least in cartography. For Merton the value implies that the results of scientific inquiry are to be made freely available; after I have achieved scientific knowledge and published my results, those results are there, free for all to use. And what could be more free than a map, there open to view? Indeed, there is an image of the map—where each of us maps a different area and we then create from the pieces, as from a jigsaw puzzle, a larger map—that is a comforting analogue to common sense discussions of progress in science. Furthermore, if today maps and geographical data can be expensive indeed, it nonetheless seems as though a map remains a map, something that once purchased is in some ways a public object, there to be seen by all.

But we know that there is another side to the matter. As I noted earlier, almost two thousand years ago Strabo held that geography "regards the activities of statesmen and commanders."[1] And we know that during the age of exploration maps were held to be national secrets, of such extraordinary importance that sea captains knew that if they lost an atlas they might as well not bother to come home. And we know, too, that maps—and geographic data—have had in this century continuing military importance; during the Cold War both United States and Soviet maps were routinely altered to make them less useful to the enemy, while the work of government mapping agencies such as the Defense Mapping Agency was highly classified.

What are we to make of this situation, of the two very different ways in which access to maps and geographic data is controlled? In what follows I shall argue that the move from manual to automated cartography and geographic data analysis has been an event of significant proportions. The introduction of large and complex systems of software and hardware has recast the authority relationships within the everyday practices of those who use geographic information systems and within the institutions in which they work. Furthermore, the increasing availability of geocoded data, in the context of means for rapid dissemination and analysis, has made those data valuable commodities, to such an extent that Merton's communalism now seems a quaint dream.

Indeed, I shall argue that in important respects the very idea of ownership that has undergirded the development of science over the last two hundred years is coming apart, replaced by a system in which, increasingly, hardware, software, and even data are leased from the "real," owners, with their uses strictly restricted. As technological developments make it increasingly important where we are on the map and not on the ground, we can no longer be said to stand on the ground. We no longer own our own location.

On the roots of the ownership of ideas in science

In order to understand the changes in the ideal of communalism associated with geographic information systems, it will be useful to consider the ways in which we typically ascribe rights of ownership in science. In effect, we live with two very different sets of justifications for ownership. On the one hand, it is typical to believe that a right to ownership can derive from one's labor. I work on something, and as a result it is, at least in part, mine. This view was codified as far back as John Locke's *Two Treatises on Government*. There he argued that

> man, by being a master of himself and proprietor of his own person and the actions or labour of it, had still in himself the great foundation of property; . . .
> Thus labour, in the beginning, gave a right of property wherever anyone was pleased to employ it upon what was common. . . . [2]

On the other hand, I can also gain a right to ownership where it is not labor so much as creativity—even an instantaneous burst of creativity—that is central, and especially where the object that I create can be seen as expressive of my very being, my personality. This is just what we expect to see in a painting, for example; the artist who works slowly gets no more credit than the one who works quickly, just because it is not the amount of work that is at issue, but rather the quality of the worker's contribution. This view, too, has been codified, in this case by Hegel in the 1821 *Philosophy of Right*. [3]

> A person must translate his freedom into an external sphere in order to exist as Idea. [4]

The rationale of property is to be found not in the satisfaction of needs but in the supersession of the pure subjectivity of personality. In his property a person exists for the first time as reason.[5]

Mental aptitudes, erudition, artistic skill, even things ecclesiastical (like sermons, masses, prayers, consecration of votive objects), inventions, and so forth become subjects of a contract, brought on to a parity, through being bought and sold, with things recognized as things.[6]

On this view an individual *becomes* an individual only through interactions with others, through existence within a community. But in order to be a real individual it is necessary that one make manifest the products of one's inner spiritual or intellectual life—and property is the way in which what is inner can become not only external, but also permanent. And it is this view that is behind the forms of property law, and especially intellectual property law, that are common to Europe.[7]

In a sense, this second view appears more in keeping with the common perception of science than does the other, labor view. In fact, the popular image of the scientist is of a person who acts on just such bursts of creativity; it is common—just watch television programs on science, or read the science press—to see the work of science as a matter of putting together disparate elements through an act of insight.

By contrast, perhaps the worst thing that can be said about a scientist is that he or she is a plodder, a person who spends day after day, night after night, collecting data. Such work is seen as too mundane to be esteemed as real science. It is not, of course, as if this sort of work is not done within science; indeed, it is absolutely essential. But it does not receive the esteem of the creative work, and in fact is very often not even seen as science. Still, it is esteemed in another way; the people who do it—the secretaries, lab technicians, research assistants, and editors whom I mentioned earlier—are paid.

I have noted that both views of rights and responsibilities are widely held; in fact, it is not inaccurate to say that both are seen and invoked as common sense. This certainly is because a version of this view is at work in everyday conceptions not only of the work of scientists, but also of work more generally. As I suggested in Chapter 5, theorists from Marx to Taylor to Braverman have pointed out that in the division of labor as it operates in our society there is a distinction between management and labor, between thinking and doing.[8]

On the idea of intellectual property

But it is important here to see that these images, of labor and creativity, the head and the body, were codified at particular moments in the history of science and of the Western economy. Before the advent of the modern factory system, an important place for the storage of human knowledge was the human body, in the form of know-how, the ability to do certain tasks. And so, a worker—a carpenter, for example—acquired an ownership right in a building by virtue of working on it, by incorporating his labor in it. But although this know-how was "stored" within the individual, it was at the same time strongly embedded in sets of social institutions.[9]

By the time of Locke, though, it became possible to think of property as something that was strictly individual. Granted, he saw the acquisition of property as something that had a broader social impact. But this was dealt with in his proviso, that one could gather property from nature only where what was left to be taken was sufficient for everyone else, where "there is enough and as good left in common for others."[10]

Although there is argument about the earliest instance of copyright, it is now agreed that current statutes exist in a direct line that began in Europe in the fifteenth and sixteenth centuries.[11] The earliest formalization of the Hegelian view, in this case derived from natural law, was in the French revolutionary laws of 1791 and 1793. This law focused directly on the rights of authors:

> Authors of writings of all kinds, composers of music, painters and designers who make engravings or drawings, shall enjoy during their entire life the exclusive right to sell, prepare for sale, and distribute their works in the territory of the Republic, and to assign the property therein in whole or in part.[12]

As I suggested above, this personality theory, more popularly referred to as the doctrine of "moral right," bases the justification of rights in intellectual property on the belief that creative works are the external embodiment of the personality of the author, and thus ought to be protected as an element of the protection of the individual.[13] Here the rights include the right to say when a work may be divulged and the right to have one's name attached to a work (or to see that it not be).

Typically—although as we shall see, not universally—the rights may not be alienated; they belong through the lifetime of the creator to the creator, and thereafter devolve to that person's heirs. They are permanent. Moreover, a person to whom an assignable portion of the right is given (a publisher, for example), may not without permission alter the work. Indeed, under certain circumstances an author may require that works, because they are no longer expressive of her personality, be retracted.

In contrast, the British law developed as a system of privileges accorded to printers, and hence as a means by which the labor of printers might be protected. In Britain this system, protecting printers and based on the assumption that authors were supported through a system of patronage, gave way to the first formal copyright law, the Act of 1709 (or, the Act of Anne).[14] And in fact, although it is common to think of copyright as applying primarily to works of creativity, in English (and soon American) law, the earliest disputes were more concerned with works of information, works in some ways like information systems, than works of "inspiration." Birrell, for example, noted that

> In reading cases in the [English] Reports for the last hundred years, you cannot overlook the literary insignificance of the contending volumes. . . . The question of literary larceny is chiefly illustrated by disputes between book-makers and rival proprietors of works of reference, *sea charts*, *Patteson's 'Roads'*, the antiquities of Magna Graecia, rival encyclopedias, *gazetteers*, guide books, cookery books, law reports, post office and trade directories, illustrated catalogues of furniture, *statistical returns*, French and German dictionaries, Poole's farce, "Who's Who?," Brewer's "Guide to Science" [emphasis added].[15]

And so, the rights to works of geography and of science more generally came to be characterized in these individualistic and economic terms, terms suitable for the mass produced work and the work produced within a system, like that of de Prony, based on the division of labor.

If rights to such published works were often seen in this way, there was another side to the story. The personality or moral right theory of property, theorized in Hegel's *Philosophy of Right* and codified into law in Western Europe, can perhaps best be seen as part of a romantic reaction to the ascendant mass production and individualism. And indeed, even in countries—such as the United States and England—where the Lockean theory was more popular, interpretations of ownership disputes in written works came to see early on the written work as having a twofold existence: as a product, but also as a repository for ideas that in a sense existed outside of it. Here, of course, the appeal was to the notions of intelligence and genius that have come to be so much common sense in work on science today.

In geographical work these ideas were, perhaps, more common than today; through the eighteenth and nineteenth centuries the heroic geographer–explorer was much in the news.[16] Indeed, the value of a degree of longitude, a matter that both captured the popular imagination—as a matter of both intellectual and physical struggle—and became an element in the most mundane of volumes, is perhaps a perfect example of the way in which, in geography, these two forms of ownership have coexisted.[17]

These two views of intellectual property lead in practice to very different systems of control. On the one hand the focus is on authorship and creation, on the other it is on economics. But they share roots in a time before the hegemony of Taylorist and Fordist forms of production, and before the development of the corporation as a legal individual. As a result, they have faced increasing pressure, in the face of those forms of production, of the increasing value of information, and of its increasing mobility. So today we see a recasting of this system of ownership. In the process the possibility of the communalism envisioned by Merton is fading, just as in many cases the scientist is no longer able to claim to be the owner of the results of her work.

New means for intellectual-property regulation

In the case of intellectual property, almost as soon as it was codified it became clear that there would be a need for means to prevent its misappropriation. Today a variety of means for its protection are used; two international conventions, the Universal Copyright Convention and the Berne Convention, are the most broadly used, but they operate against the background of a range of other actors.

In Western Europe, three groups, the European Commission of the European Community, the Council of Europe, and the Organization for Economic Cooperation and Development (OECD) are also active. Finally, and controversially, the US has been active in promoting its own views of intellectual property through the General Agreement on Tariffs and Trade (GATT) and its successor, the World Trade Organization (WTO), and through bilateral trade agreements. In effect, other countries have been faced with a need to develop intellectual property laws that will meet the desires of a wide range of powerful interests. In each case the issues of computer software and databases have been absolutely central— sometimes to the exclusion of all other interests.

The roots of the current situation lie in the late nineteenth century, when the protection of intellectual property was codified through a series of international conventions. For written works the central one was the Berne Convention, developed in a series of meetings from 1884 to 1886, and codified in that year. It has since been revised a number of times, and each revision is generally more strict than the last.

The details of the current version of the convention need not detain us here, other than to note that beyond the expected listing of protected works, descriptions of the length of copyright protection, and specification of the means by which each country ought to deal with foreign works, it defines the works to be protected as those that are "literary and artistic" and stipulates that the author of a work has not only an economic, but also a moral right. Hence, the Convention draws together elements of Anglo-American labor-based property theories with those of the French, personality-based theory.

All of the signatories of the Berne Convention (including all of the countries of Western Europe, all of Central and Eastern Europe (except Albania), and the United States (which belatedly signed the Convention in 1989) are therefore committed to a view of copyright that sees it as a matter of the protection of creative works, and all are thereby required to enact legislation that protects both the economic and the moral rights of authors. But there have in the last few years emerged a set of alternative means for the protection of intellectual property, means employed primarily in the developed nations of the West, and means that attempt to provide protection for works that do not fall under the usual copyright protections of works that are creative and non-useful.

One of these was developed by the European Community.[18] The Commission of the European Communities has since 1991 published two directives, one relating to software and the other to databases. The first of these purports to be merely an interpretation of the Berne Convention, although it does have one important point for current purposes; it appears to allow for the copyrighting of the "look and feel" of software.

The second, and more important, establishes means for the protection of databases. In fact, if there had been something odd about copyrighting software, the thought of copyrighting databases seemed strange indeed. They are, after all, almost inevitably the works of groups of people, anonymous, constantly changing, never fully viewed, and perhaps most of all, works not of inspiration but of the utmost drudgery. Most existing laws, to the extent that they treat them at all, do so as compilations. But the question arises, compilations of what? To what extent, after all, does a list of baseball scores resemble a Norton Anthology? And laws had varied substantially, with some requiring a great deal of originality, and some requiring none.

The Commission's action, surely, was motivated in part by a US court case, *Feist* v. *Rural Telephone*, in which the Supreme Court had argued finally and definitively, after two centuries of ambivalence, that labor is simply not enough, and that works simply of labor may not be protected under copyright statutes.[19] The Commission was clearly taken with the idea of databases; it called them the "hypermarkets" of the future, compared them to raw materials, and declared that compiling a database is like making a collage.

The Commission dealt with the issue of originality by creating two levels of protection for databases. First, to the extent that they are the result of originality, in selection or arrangement, they may be protected under standard copyright law. Here, the less selection has occurred, the more arrangement is required, and vice versa. Second, to the extent that they fail to meet this standard they may be protected under a second, *sui generis*, form of protection, that proscribes the "unfair extraction" of parts of a database.

The Commission's actions went against the desires of the international intellectual property organizations, which have—perhaps in the interests of maintaining their own authority—persisted in resisting attempts to create new forms of intellectual property. But a series of other actions have gone well beyond the Council and the Commission, and have created dramatic new means for the definition and protection of such property. Here, in large measure because of its size, the US has been at the forefront in using creative means not just for going beyond the Berne Convention, but for actually replacing it. This has involved the development of what has been called a "menu" of options for dealing with intellectual property issues, and especially those of software and databases.[20] This is the primary way in which the US has dealt with Central and Eastern Europe.

The US has actually used seven strategies to exercise control over intellectual property in the international arena. First, it can prevent the importing of goods by claiming unfair trade practices, under Section 337 of the 1988 Trade Act, that made those claims easier to press by softening the rules of evidence. Second, in the case of goods already imported, it can file suit against the importers and distributors. Third, it can threaten sanctions under Section 301 of the Trade Act of 1974. Fourth, it can threaten countries with the loss of GSP (Generalized System of Privileges) under the Trade and Tariff Act of 1984. Fifth, it can offer technical assistance to countries that comply with its desires. Sixth, it can invoke strict reciprocity requirements, allowing trade with other countries only where they establish standards of protection equal to those in force in the US. And finally, it can support multilateral trade agreements, through GATT and the WTO, containing intellectual property restrictions.

GATT, established in 1947, was originally conceived as a mechanism for removing "distortions" in international trade. But as the recently completed Uruguay Round of talks started in the late 1980s, the US pressed a new strategy, linking trade with intellectual property. The US believed that the World International Property Organization (WIPO) had been too much influenced by developing nations, and not firm enough in preventing piracy, and it also believed that the Berne Convention was too lax in allowing for systems of compulsory licensing. It promoted the GATT strategy as one of more benefit to advanced, intellectual property-exporting nations. It hoped, under the aegis of GATT, to establish support for its understanding of intellectual property and to rely on GATT's system of dealing with nullifications and impairments, that is, with violations.

And indeed, when the treaty was signed in 1993, it included a stipulation that:

1. Computer programs, whether in source or object code, shall be protected as literary works under the Berne Convention (1971).

2. Compilations of data or other material, whether in machine readable or other form, which by reason of selection or arrangement of their contents constitute intellectual creations shall be protected as such. Such protection, which shall not extend to the data or material itself, shall be without prejudice to any copyright subsisting in the data or material itself.[21]

At the same time, the US has been developing a series of bilateral trade agreements directly with other countries. The copyright statutes there are either newly created or in the process of creation, and the US has taken the opportunity to encourage these countries to adopt intellectual property systems consistent with its own desires. For example, a side letter to its 1990 Business and Economic Treaty with Poland stated that

The Republic of Poland will explicitly extend copyright protection to computer software before 1 January 1991. The terms of protection for computer software will be equivalent to that provided for other literary works.[22]

This, in fact, is precisely what has happened, as Poland in December 1993 passed a new copyright statute which is very much in the American mold.

Similar stipulations were included in an agreement with the Czech and Slovak Republics, where a 1990 trade agreement stated that "each Party shall, inter alia, provide copyright protection for computer program and databases as literary works under its copyright laws." And a side letter to the 1992 Romanian trade agreement stipulated that

Each Party shall protect . . . all types of computer programs (including application programs and operating systems) expressed in any language, whether in source or object form which shall be protected as literary works; and, collections or compilations of protected or unprotected material or data whether in print, machine readable or any other medium, including data bases, which shall be protected in so far as they constitute an intellectual creation by reason of the selection, coordination, or arrangement of their contents.[23]

Similar agreements are found in trade agreements and side letters with other nations, such as Bulgaria and Albania.

This appears, under US prompting, to be the trend of the future. The US has substantially different interests in intellectual property restrictions than do less developed countries. They prefer weaker standards, as means for promoting their own development. By contrast, the United States, having already become developed and relying increasingly on exports containing large proportions of intellectual property, prefers strict standards.

And what this means is not as simple as it may seem. For ultimately the ways in which intellectual property is protected have fundamental impacts on the nature of work. The advocates of a moral-right view of intellectual property have traditionally attempted to maintain an approach in which the center of creative activity is the individual, but an individual within a community. There the individual receives not just economic but also social and cultural benefit from the process of production.

Yet the United States, in its promotion of the extension of the moral-right view to databases, has pointed in a very different direction. For much of the data in question is in the hands of large companies, of media conglomerates like Time-Warner and Sony, and of the direct marketing industry. And so, the organization of labor and of responsibility is very different.

In the traditional image, the one underlying both the labor and personality theories of intellectual property regulation, the individual creator or laborer was at the center. The purpose of regulation was to support that individual, whether a printer or a painter. The two systems treated the issue of the alienation of works very differently; in the labor theory a work sold was one gone, where in the moral right theory the creator always held a measure of control over the work. But in either case, there was a sense that the creator and seller of a piece of intellectual property, whether a painting or a book, was in a way very much like the buyer. Indeed, having become a buyer a person then became a potential seller. In the Newtonian world of the Lockean property theory, the image was that of the flow of goods in a kind of economic space. But if in the Hegelian world the image was more of the creator using property to create his or her personality, it remained that the creator and the viewer or buyer shared the need to engage in that process. So in either case, we were all buyers and sellers.

But the migration of data to large corporations, in a context in which a corporation can own data in the same ways as can an individual, has meant that this parity has been lost. Indeed, if today in the areas of data and software the creator and the buyer are alike, they are alike in that in many cases neither holds a right to ownership. The creator is a laborer, who lacks a moral right to works, and who can better be seen as working within a contractual situation in which what is being remunerated is labor, and not the fruits of that labor. At the same time, users of data and software very often have no ownership right to either. Rather, they merely license both data and software, paying for the ability to use them in a prescribed way, in a prescribed place, for a prescribed period.

Who owns geographic information systems?

In what sense is this relevant to the use of geographic information systems? And how is it related to the issue of communalism? The answer is in part connected to the fact that science, and Mertonian science, evolved alongside the intellectual property system. In a free-market system of the sort that has existed in the West, the very idea of communalism—which on the face of it seems contrary to the economic free market—relied upon a way of supporting the means for the dissemination of research results. To give one sort of credit to those who created the results and another to those who disseminated them was under those circumstances an ideal compromise.

Yet for that compromise to work, it required that scientists be able to represent themselves as autonomous agents, able to know the works of the systems in which they operated from beginning to end. The scientist, at least in principle, needed to be able to vouch for the quality of the data collection as much as for the quality of the representation of those data in final, public form. But because the creation of geographic information systems has involved the corporation as a major actor, it substantially altered the ability of scientists and other users convincingly to argue that they have that knowledge.

And as I have noted elsewhere,[24] a geographic information system in fact consists of a series of actors—those associated with hardware, software and data. Where once upon a time—and this is of course still true in many areas—the individual scientist might build the necessary apparatus for data collection and analysis, collect the data, and analyze them, in the case of a typical geographic information system these three tasks are divided among at least three actors. There are those who design, build, sell, and maintain hardware; there are those who design software for data collection, analysis, and representation, and perhaps do the analysis; and there are those who collect and maintain the data.

If, as we have seen, this division of labor is nothing new (indeed, William Zilsel notes that it was only in the time of Galileo that it became acceptable for a scientist both to collect data and develop theories; in the Middle Ages these two tasks had been rather strictly separated),[25] what is new is the following.

Where the traditional means of legally defining the ownership of scientific work—the copyright and the patent—were explicitly designed to support and enhance the intercourse in ideas, the corporation has been able to introduce a new set of ways of controlling its property. In the case of software the most obvious of these is trade secret law. Indeed, in most commercial software packages for statistical analysis and geographic information systems the algo-rithms that are used for data analysis and representation are in some respects protected by trade secret law, and hence not open to outsiders.

If there are problems with software, there are equally problems with data. Providers from the United States Census Bureau to private corporation—for different reasons—today exert considerable control over large-scale databases. They may both establish and maintain the standards in accordance with which data are collected and stored; at the same time, they are able to control the avail-ability of both the data and the information necessary for an adequate assessment of those data. And by controlling access, they of course control the ability of the user of the system to treat those data as potentially communal property.

If these cases—of software and data—apply more generally to those using large information systems, there is an area in which the issue of communalism arises that is distinctly geographical. If we turn to the history of the determination of location, we find significant battles over technology. Here, of course, the attempts to develop technological means for the determination of longitude come to mind, as do attempts to measure latitude.

But once those technologies were in place, they were widely used. Indeed, as we have seen in the cases of de Prony and Birrell, there was a huge market for statistical and navigational tables, just as the clock, the key to the determination of longitude, quickly became a readily available, if costly, commodity. Hence, once the system was established, anyone—at least in principle—had access to informa-tion about location.

But the development of geographic information systems and now global posi-tioning systems has transformed the situation, and in a strange way. Global positioning systems have become increasingly popular. Now quite inexpensive, they are widely used, not only in traditional areas like surveying, but also in outdoor sports. A quick look through any sailing, fishing, or camping magazine or catalog will reveal an array of equipment and accounts of how to use it.

But the location that is displayed in the system is not the "real" location on the ground. Any system, whatever its accuracy, represents a fixed object as drifting over time. The drift is, of course, on the map; it is in the data. Now, this would matter not a bit, except for a single fact; decisions are made not on the basis of where one is, but rather on the basis of where one is on the map. As Laura Kurgen put it,

> The older and perennial question of "Where am I?," the question that gives rise both to panic and to new discoveries, has been replaced or displaced by a stranger interrogative, "Which pixel am I standing on?"

The question is no longer "'where am I' on the earth, but where am I on the map?"[26] Where the global positioning systems consist of private satellites, privately produced hardware, and licensed data and software, Kurgen might have added the question "Where am I on *their* map?" Because in such a situation we no longer have maps of our own, maps to give away.

7

THE DIGITAL INDIVIDUAL IN A VISIBLE WORLD

We have seen by now that the place of geographic information systems is complex. They are limited in the ways in which they can represent peoples, places, and nature. Their use is associated with a refiguration of the practice of science. And because, in part, that refiguration is expressed in changing conceptions of ownership of the tools and products of science, the systems are associated with a set of broader social changes.

There is another way in which the systems are associated with broader social changes, and social changes that are profound indeed. This is in the area of privacy. If geographic information systems have been used outside of the academy for as long as they have existed, the last few years have seen an explosion in that use. Central to that increase in use have been geodemographic systems, commercial offshoots of geographic information systems used primarily by the direct marketing industry, but also used in intelligent transportation systems, banking, the wireless communication industry, and of course in direct marketing and site location.[1] When these systems are connected to global positioning systems, and the two to satellite or other remotely-sensed imagery, one has created a system of great power, and of great utility for the storage and analysis of information and for extended surveillance on individuals and groups.

For that reason the systems raise important questions about the nature of privacy. In fact—and just because of their use of geocoding—the systems erode the utility of traditional tools of privacy protection. But at the same time, they raise deeper concerns. They are associated with an erosion of the traditional forms of the private and the public—and as a consequence require that scholars, activists, and the public at large rethink both realms.

On the genesis of the right to privacy

The formulation of the explicit right to privacy has its origins in urban society in the late 1800s. Prior to that time there was no well enunciated privacy right; certainly people had privacy, but it was guaranteed by the existence of a landscape within which certain practices could by and large be expected to be private.[2] But the economic and technological transformation of that landscape in the late nineteenth century solidified the distinction between the urban and the rural. And in doing so, it set the stage for the formulation of privacy as a right.

The urban, as Georg Simmel bemoaned in his famous "The metropolis and mental life" and Louis Wirth later formalized for American urban sociology, became a place in which an individual could choose to remain isolated and anonymous.[3] The problem for privacy created in this new landscape, where the actions of the individual seemed so little constrained by custom, was solved by the development of a formalized set of privacy guarantees. In the United States they were set out in a famous law review article near the end of the last century, where Warren and Brandeis asserted that the individual has a right to privacy, where that is the right "to be left alone."[4]

If it has been common in thinking about the individual, at least since the 1650s, to imagine the individual as having a single, coherent identity, there is a sense in which popular practice during the era after Warren and Brandeis took a very different tack, one more consistent with the views set out by Simmel. Not only was the modern urban individual seen as living a fragmented existence, that fragmentation was seen as beneficial, and as something that needed to be supported.

The idea of privacy is a complex one; it has multiple meanings and is supported by multiple justifications. In an overview of the related issue of data protection, David Flaherty describes privacy as involving all or some of the following:

The right to individual autonomy
The right to be left alone
The right to a private life
The right to control information about oneself
The right to limit accessibility
The right of exclusive control of access to private realms
The right to minimize intrusiveness
The right to expect confidentiality
The right to enjoy solitude
The right to enjoy intimacy
The right to enjoy anonymity
The right to enjoy reserve
The right to secrecy[5]

It should be clear that this list incorporates elements of an extremely complex understanding of the individual and of the relationship between the individual and the community. The right to autonomy and the right to be left alone, for example, are almost at opposite ends of the spectrum; one involves a right to engage in motivated action, while the other involves the right not to be the object of the motivated action of another. Similarly, while the right to intimacy may be uncontroversial, the rights to anonymity and secrecy certainly are.

Justifications of the right to privacy usually derive from the notion of the individual as the owner of his or her self. There the right to privacy involves a right to incorporate into that self a certain amount of information, and the definition of the right involves the setting of the amount that is to be protected. Other justifications rest more directly on a notion of human dignity, where certain features of an individual's life are taken simply to be privileged.

But these justifications in turn rest upon one or more of several deeper notions *of* the self, and of the individual, and these notions are not always made explicit. As Herbert Fingarette has suggested, one such notion, and one prevalent in the twentieth century, sees human life as a matter of self-creation; here—existentialists have been proponents of this view—there is no individual outside of this act of creation. Another view of the individual is exemplified in Freud, where human life comes to be seen as a matter of self-discovery, of finding out what is there. Finally, there is a telic view, deriving from Aristotle and then from Hegel and Marx, that sees human life primarily as a matter of self-fulfillment, of becoming what was always potentially there[6] (see Table 7.1 for more detail.) Each of these three views is closely related to a way of understanding the justification for privacy, but the relationship is not simple, because the notions of the individual are, in turn, related to evaluations of the relative value of the individual and society. Moreover, each is associated with different areas of society, and in a complex society any individual is likely in different contexts to appeal to different sets of justifications for the maintenance of a right to privacy. Hence Flaherty's list, which might at first seem to reflect a systematic inconsistency in the concept of privacy, but which in fact is simply a reflection of the complexity of society.

And so, for those who see the individual as something to be created—and who value the individual—the right to privacy becomes an absolute right, just because it is the means by which one is given the opportunity to decide the nature of the public face that will be revealed. Here the public and the private are equally creations of the individual, who has a right to control both. Yet for those who devalue the individual, seeing societies as best created through joint efforts, privacy becomes an impediment to that end.

In contrast, those within the telic tradition who value the individual often argue that the development of the individual, the fulfilling of an essence, needs to be allowed to proceed unhindered. And those within that tradition who favor the

Table 7.1 The idea of privacy

| Mode of analysis | Tradition | Primary center of value | | | |
| | | Individual | | Social | |
		Evaluation of privacy	Goal of privacy regulation	Evaluation of privacy	Goal of privacy regulation
Philosophical	Existentialist	Privacy supports construction of the self	Supporting the autonomy of the individual	Privacy an impediment to the creation of the good society	Supporting the autonomy of the social
Psychological	Self-creation				
Cultural	The arts; science				
Philosophical	Aristotelian	Privacy prevents impediments to self-fulfillment	Supporting the autonomy of the individual	Privacy destroys community	Supporting the autonomy of the social
Psychological	Self-fulfillment				
Cultural	Marxism; capitalism				
Philosophical	Freudian	Privacy a means of controlling access by others to the self	Negotiating the amount of information that ought to be available	Privacy a protection of deviance	Negotiating the amount of information that ought to be available
Psychological	Self-discovery				
Cultural	Modernism				

communal over the individual tend to argue that the development of the community needs to be given priority; here the needs of the community come to be identified with those characteristics not of the actual, empirical individual, but rather of an ideal individual, discernible only by those who have merged their interests with those of the community. Here, Hegelians argue that the individual right of privacy tends to be destructive to the community.

Indeed, it is no surprise that in Marx's turn on Hegel, where property becomes suspect, privacy is not a real value. This is what we see in the countries of Central and Eastern Europe, where the devaluing of privacy was a central feature of political ideology under Communist rule. Many would of course say that under Communist rule there was no right of privacy, and that the concept simply disappeared. But the above analysis suggests that it is more accurate to say that in those countries notions of the relation of the individual to the social, and of the visible to the invisible, underlie a view wherein the dominant ideology sees the existence of the hidden individual as less a source of hope than a sign of failure, and where other notions of and justifications for a right to privacy are rendered obscure—but not effaced.

Finally, in the modernist tradition that sees the "real individual" as something hidden inside, waiting to be discovered, those who value the individual tend also to value a right of privacy, as a means for the control by the owner of that self. Here restrictions need to be placed on the right of one individual to dig too deeply into another's life, to see the hidden core that constitutes the true self, which is fundamentally private. By contrast, those who devalue the individual in relation to the group tend to see this same hiddenness as no more than an opportunity for the true self, as a potential source of deviance and disorder, to remain beyond the view of the group. The essential thing to see here is that on this view there can be *degrees* of hiddenness. The right to privacy is not absolute, but rather is a matter of the establishment of a boundary between that which can be revealed and that which ought not to be revealed. The boundary can be negotiated.

On the value of privacy

As we shall see, in the United States most of the recent privacy-related cases that the Supreme Court has decided have been cases that concern activities—like drug smuggling (but also terrorism and child pornography, together often invoked as a kind of mantra in computer-network-related cases)—that have a sordid side. Few in the public are galvanized to protest by the sight of a drug smuggler being sent off to prison.

And indeed, for this very reason many people simply shrug off the changes that have been occurring in the legal definition of the right to privacy, and hence in the institutionalized definition of the private realm itself. Further, some, seeing themselves as hard-headed, have argued that the appeal to the private is simply nostalgic, a sign of a kind of mental illness. And finally, some have argued that since any privacy regulation is certain to work to the benefit of the wealthy—one (apparently) apocryphal story has it that the original Warren and Brandeis article grew out of their distaste for the paparazzi—we ought just to give it up, to live lives that are transparent to all.

Yet a bit of reflection suggests the difficulty with all of these positions. The difficulty is, putting the matter simply, that the private realm performs important functions in the life of the individual and the group. It is in private that people have the opportunity to become individuals in the sense that we think of the term. People, after all, become individuals in the public realm just by selectively making public certain things about themselves. Whether this is a matter of being selective about one's religious or political views, work history, education, income or complexion, the important point is this: In a complex society people adjust their public identities in ways that they believe best, and they develop those identities in more private settings.

As scholars we are all well aware of this. Few of us, after all, would wish every draft of every paper and lecture to be open for public scrutiny. And we are aware of it as political beings; in a society in which political power is unevenly distributed the possibility of the less powerful becoming more powerful depends on the possibility of private activity.

The ability to engage in these adjustments is contingent on the possibility of there being private places in which the elements of the identity can be assembled and tried out. And it is contingent on there being some realm of private data. Just as we rely on the possibility of doing things in places that are out of sight, we rely on the possibility of some facts about us drifting out of sight after the passage of time. We all assume that there are things about us that others will forget, and we are thereby able to feel that we live in a society where there is the possibility of redemption.[7]

What I have said of the individual applies, too, to the group. The creation of a group is contingent on the development of a border, a boundary between the member and the non-member, the included and the excluded. If in American society the discourse about the privacy of groups is less well theorized—indeed, to most the idea of "group privacy" seems almost an oxymoron—the idea is nonetheless embedded in everyday life. Political and religious groups wish to plan and pray in private, and even places wish to have some sense of control over the image that is propagated of them.

The potential refiguration of what counts as a right to privacy, then, has the potential of having important consequences at a variety of scales. So in the context of a general acceptance by jurists—not to mention the computer software and hardware industries—of the inevitability of the technological change that feeding is the redefinition of the right to privacy, it is all the more important to understand the potential implications of these changes. As I shall argue, geographic information systems and geodemographic systems have the potential to effect changes in this right in a number of ways.

Regulating privacy

Within the context of attempts to establish guidelines for data protection the modernist view, the one dominant in the United States and in Western Europe, has led to the specification of a series of features expected to render the appropriate level of protection. Typically—these are now widely seen as constituting the core of "Fair Information Practices"—it is argued that: (1) the existence of a system of information must be publicly available knowledge; (2) individuals ought to have access to data about them, and ought to be able to correct erroneous information; (3) personal data ought to be collected only where necessary; (4) personal data ought to be used only for the purposes for which they were collected; (5) personal data ought not to be disclosed to another group or agency without some sort of consent; and (6) personal data ought to be protected.[8]

The codification of these standards responds to fears expressed in a large number of books published beginning in the 1970s, books with ominous names that referred to "the surveillance society" and the like.[9] At its root is the notion that the existence of computers has led to a situation in which knowledge will mean control, and individuals will lose their abilities to be themselves. Hence, the rules attempt to do the following:

- assuage fears of a surveillance society, by requiring that the existence of sets of data be publicly known;
- assuage fears that an individual will be wrongly labeled; and
- assuage fears that individual pieces of data will be collected and combined into a dossier.

And the concern is less with the prevention of the collection of data about individuals than about the rendering consistent and visible to each individual the types of data that are held, where they are held, and by whom.

At the same time, though, what we see here is an attempt to establish a negotiated boundary between the individual and the social, and this is typical of the data protection statutes passed in the US and in Western Europe. It is also typical of the guidelines developed by the Council of Europe, that will become standards for statutes passed in Central and Eastern Europe.

Information systems and the right to privacy in Europe

In Europe, the issue of privacy has been under constant discussion since the late 1960s.[10] In fact, although it had been a matter of concern since the 1940s, under the rubric of human rights, it was the development in the 1960s of computer technology that brought the issue to the table in the form of a more immediate and dramatic issue. That period was one in which the images of Big Brother and of an Orwellian *1984* seemed all the more pressing, as large-scale computers appeared to make possible the development of government-run databanks, containing dossiers on every individual within a country. And so, by the late 1970s the OECD, the Council of Europe, and the European Commission, had done substantial work on guidelines for the regulation of the use of personal data within the context of the computer.[11] In the intervening years these earliest reports have been joined by a series of reports, recommendations, and guidelines, some binding and some not. Today they constitute a set of regulations, directed in the first instance at Western Europe, that define the kinds of personal data that can be freely transmitted, the kinds of restrictions that can be placed on the flow of personal data, and the rights of individuals to control those data.

Although these standards were in the first instance developed in Western Europe, their reach has been much greater. The European Community consists only of Western European nations, but its standards have been designed to mesh with those of the Council of Europe, which includes fifteen other European countries. And both, in turn, have developed alongside standards developed by the OECD, which includes the US, Japan, Canada, Australia, and New Zealand. As a result, and because included are all of the world's major computer users, the standards can be seen as in some sense universal.

The earliest large-scale discussions of data protection in Europe occurred in the Council of Europe, at about the same time that such discussions were beginning in the United States, and the Council's documents have been important elements in subsequent documents produced by other bodies. Established in 1949, the Council has as its mandate the setting of standards across Europe. These standards, once ratified by an individual country and written into its laws, become binding on that country. In 1971 the Council appointed a Committee of Experts to study data protection in the private and public sectors, and the deliberations of that committee were published in 1973 and 1974.[12]

The Council adopted a system of fair information practices, recommending that individual member states adopt regulations requiring that: (1) information be accurate and up to date; (2) information be relevant to the purposes for which it was collected; (3) information be legally obtained; (4) information be kept for a specific period; (5) information be used only for the purpose for which it was collected; (6) individuals have the right to know what information is kept, and to whom it is disseminated; (7) old information be destroyed; (8) security measures be kept; (9) only those with a valid reason get access to materials; and (10) statistical materials be released only in aggregate form.

These principles were built into the Council's final convention, published in 1980.[13] There is, though, one significant difference between the two documents; the final version deals in some detail with the issue of transborder data flows, an issue missing from the earlier document. It declares that

> [T]here shall not be permitted between Contracting States obstacles to transborder data flows in the form of prohibitions or special authorisations of data transfers. The rationale for this provision is that all Contracting States, having subscribed to the common core of data protection provisions . . . offer a certain minimum level of protection.[14]

Like the Council's final convention, the work of the OECD (which was begun after the publication of the Council's recommendations) was promulgated in order to

> apply to personal data, whether in the public or private sectors, which, because of their nature or the context in which they are used, pose a danger to privacy and individual liberties.[15]

And like the Council, and under pressure by the argument of the US (a member country) that restrictions of transborder flows of information were both inconsistent with freedom of speech and unwarranted restrictions on free trade, the OECD took a strong stand on those flows, arguing that:

> Member countries should take all reasonable and appropriate steps to ensure that transborder flows of personal data, including transit through a Member country, are uninterrupted and secure.

> A Member country should refrain from restricting transborder flows of personal data between itself and another Member country unless the latter does not yet substantially observe these Guidelines.

> Member countries should also ensure that procedures for transborder flows of personal data and for the protection of privacy and individual liberties are simple and compatible with those of other Member countries which comply with these Guidelines.[16]

In the case both of the Council and of the OECD, individual countries were given the opportunity to develop more restrictive regulations concerning individual elements of privacy regulations. For example, a country might restrict the dissemination of information on magazine subscriptions, on the basis that such data may reveal the political or sexual preferences of subscribers. Nonetheless, both groups strongly linked the issues of privacy and of transborder data flows, and argued that a solution to any incompatibility needed to be based on an easing of privacy restrictions to meet the needs of data flows.

By contrast, in an early draft the European Commission sent shock waves through the computer and business communities by arguing that the solution lay in the other direction, by creating a law that combined the strongest elements of the privacy laws of the member states.[17] This was especially true in the matter of the individual's consent. The Directive modeled its rules on those that apply to research subjects, and stipulated that before data about an individual could be released, that individual had to give consent. The consent had to be "informed;" the individual had to be informed of the nature of the data, the purposes for which it was being used, and the owner of the data. And the consent could be withdrawn at any time.[18]

Furthermore, it stipulated that data subjects had the right

> Not to be subject to an administrative or private decision involving an assessment of his conduct which has as its sole basis the automatic processing of personal data defining his profile or personality.[19]

At the same time, lest this seem like a document that leaned in every respect toward individual privacy, a wide range of exceptions were included for public data use, including national security, defense, criminal proceedings, public safety, monitoring and inspection, and "a duly established paramount economic and financial interest of a Member State or of the European Communities."[20]

Still, the response from the business community to the 1990 Draft Directive was immediate and loud. They argued that the implementation of the Directive would cripple businesses operating in Member States, and because of the restrictions on transborder data flows would isolate the Community. As a result, in 1992 the European Parliament passed 140 amendments to the Draft Directive, creating a revised document (known as the "1992 Draft Directive") more mild, albeit still rather more strict, than the ones passed by the Council and the OECD.[21] Perhaps most notable was the rejection of the "research subject" model and the adoption of a very different version, wherein people are given not a choice to "opt in" to the use of data about them, but only the opportunity to "opt out."

Yet this is not the end of the story, for in 1995 the European Union finally passed a data protection directive that reasserted the difference between Europe and the rest of the world. The Directive required that "Member States shall provide that the transfer to a third country of personal data . . . may take place only if . . . the third country in question ensures an adequate level of protection."[22] Aimed in large measure at the United States, this Directive appears to force the United States—and thereby American companies—to strengthen its privacy protection regulations. At the same time, as we shall see, the United States has been firmly opposed to such changes, and has attempted in a range of ways to block them.

And so, in Europe, there have been a complex set of developments. In Central and Eastern Europe the regulation of data has developed against the background of two views—one American, and focusing centrally on individualism, and one developed under Communism, where the focus is on the interests of the group. Here, ironically, the view of privacy promoted by those who wish to maximize the economic values of the products that they produce through less restrictive controls on trans-border data flows has turned out in some respects to be consistent with that Eastern European view. In both, the values of the group have come to be viewed as consistent, even identical, with the values of the economy. At the same time, they have been developed in terms of a very different view, one that sees the boundary between the self and the world as something to be negotiated in terms of a wide range of social and individual interests.

The key to these codifications appears to have been the protection of the perceived interests of Western corporations. But in stark contrast to the justifications used for trade barriers in the case of agricultural and other industrial commodities, in the case of information systems what has been especially important has been the ability of advocates to invoke an image, of the unquestioned value of free trade, that has come to overwhelm other guiding images.

One element of this recast view of privacy is, ironically, a telic notion very much like the discarded Marxist one. Here society fulfills its potential where a maximized flow of saleable information allows business to flourish. But alongside this view has developed another, a decidedly modernist view. Here the central concern is one of balancing the privacy rights of individuals with the information needs of businesses and the larger corporate need of society. Indeed, here the concern with autonomy disappears; the focus is primarily informational. A right to privacy is necessary to insure that a person will be able to engage in the process of self-discovery. So privacy means data protection, and that means the establishment of a contingently placed boundary between the individual and the social; the process is one wherein the needs of the individual are balanced against the needs of that beyond the individual—the family, corporation, state—and a boundary negotiated.

As a result, though, the scales tip in favor of the computer user. For where a database is very large and the risks to an individual very small, the costs to the business of complying with strict data protection statutes far outweigh the risks to any single individual, and probably to any vocal or politically active group of individuals. Indeed, the more extensive data networks become, and hence the higher the costs and the more distant the possible risks, the more the balance is tipped in this direction.

But at the same time, in Western Europe the story has been complex. What began as a response to claims that the development of information systems would involve the violation of the individual's "liberties" or ability to engage in "conduct" has been recast in response to counterclaims that personal information is of value as much to individuals as to businesses and governments. But there has at the same time been a continuing and in some ways strengthening trend toward an approach that not only strengthens the privacy rights of individuals, but also tends, for better or worse, to increase the sense of identity of the region, in much the same way that we saw in the case of intellectual property.

The right to privacy in the United States

The story of privacy protection in the United States has been very different. In Europe, and especially Central and Eastern Europe, we see a strong corporate influence on the promulgation of privacy regulations. Nonetheless, within the European Union those regulations remain substantially stronger than the regulations within the United States, to the extent that American corporations have complained that they constitute an impediment to trade. In the United States, though, the development of data protection regulations has been hindered by a strong anti-regulation tradition, and by the continued reliance within the legal system on the view that damages to an individual's right to privacy can and should be handled in civil courts, through the system of torts. Still, outside of the direct confines of data protection, there have been important developments in the regulation of privacy, and developments that are directly connected with geographic information technologies.

Written one hundred years before Warren and Brandeis, the US Constitution does not itself enunciate a fundamental right to privacy. It does, though, in various places lay out what many scholars have regarded as elements of such a right. The Fourth Amendment, concerning searches and seizures, is one such element. It states that

> The right of the people to be secure in their persons, houses, papers, and effects, against unreasonable searches and seizures, shall not be violated, and no Warrants shall issue, but upon probable cause, supported by Oath or affirmation, and particularly describing the place to be searched, and the persons or things to be seized.

This amendment limits the actions of government. It does so by stipulating that there are certain cases in which a government (typically in the form of the police) may not without first obtaining the assent of a putatively more neutral third party (typically in the form of the judiciary) engage in searches or take the fruits of such searches. Indeed, it *defines* searches as those cases in which such assent, in the form of a warrant, is necessary.

With respect to the genesis of this amendment, one can tell a long story, as David Flaherty has, but there is general agreement about a shorter version of that story.[23] On this view the amendment arose from a perception of abuses by the British government, particularly its practice of issuing unrestricted warrants. The right to privacy since that time has been closely connected to the ability to restrict the sorts of information that is publicly available. Finally, privacy has continued to be seen as having important social functions; it is commonly seen as essential to the spiritual and personal development of the individual and the family, and hence as indirectly important to society more broadly.

Technological change and the changing right to privacy

At the time of the writing of the Constitution the house was seen as the central locus of intimate activities, and hence as the place where the intervention of the government needed the strongest justification. But if the house itself was important, in the United States it was assumed—in keeping with common law—that what was important was less that area bounded by four walls than the somewhat larger area within which the intimate activities of everyday life took place; this "curtilage" consisted of (as the *Oxford English Dictionary* put it):

> A small court, yard, garth, or piece of ground attached to a dwelling-
> house, and forming one enclosure with it, or so regarded by the law; the
> area attached to and containing a dwelling-house and its out-buildings.

As late as 1924 Justice Holmes alluded to the curtilage in *Hester* v. *United States*; turning for authority to Blackstone's *Commentaries*, he distinguished between the dwelling-house and curtilage, and the "open fields" beyond, claiming the distinction to be "as old as the common law."[24]

But a variety of social, economic and technological changes have over the last one hundred years seemed to widen the arena within which the presumption of a right to privacy ought to operate. Over this period the area within which private activities can take place has been extended beyond the home and curtilage, to the workplace, the automobile, and even the telephone booth. Yet these extensions have not been simple, nor uncontested. Some have held that when we apply the Fourth Amendment to new circumstances we need to ask the following question:

Are the actions of the government here physically like those used in making a "traditional" search of a house? Others have asserted that we need to ask a different and more encompassing question: If we were to deny that a telephone booth, for example, can be a place within which people have a right to privacy, what would be the impact of this denial on the values that the Constitution was designed to protect? Some, then, have believed that the essential issue with respect to the actions of government is the means used, while others have argued that we need to look at the ends that we—and the Constitution—seek.[25]

It is useful to focus the discussion of privacy on recent court decisions, and especially decisions of the US Supreme Court. The courts do have an impact, but, more importantly, the Court's thinking on this issue can be seen as mirroring views more widely held. Indeed, I shall argue that recent Court decisions reflect conceptions of culture and of technological change that are in accord with those implicit in practitioners' thinking about geodemographics and geographic information systems.

In fact, over the last one hundred years the Supreme Court has wavered between these two poles, for a time appealing to an ends-based view, then a means-based view, and sometimes claiming to take one view while appearing to take the other.[26] On one side we find a trend begun in *Boyd* v. *United States*, which maintained that a person's fourth amendment right could be violated even in the absence of a physical search:

> It is not the breaking of his doors, and the rummaging of his drawers, that constitutes the offence; but it is the invasion of his indefeasible right of personal security, personal liberty and private property. . . . [27]

What are important are not the means used, but rather the values or ends that might be violated by the action of the police.

On the other side is the view taken in *Olmstead* v. *United States*—a telephone tapping case—which held that

> By the invention of the telephone, fifty years ago, and its application for the purpose of extending communications, one can talk with another at a far distant place. The language of the Amendment can not be extended and expanded to include telephone wires reaching to the whole world from the defendant's house or office. . . . The reasonable view is that one who installs in his house a telephone instrument with connecting wires intends to project his voice to those quite outside, and that the wires beyond his house and messages while passing over them are not within the protection of the Fourth Amendment. . . . [28]

On this means-based interpretation a telephone tap involves no invasion of privacy, since there has not been

> an official search and seizure of his person, or such a seizure of his papers or his tangible material effects, or an actual physical invasion of his house "or curtilage" for the purpose of making a seizure.
>
> We think therefore that the wire tapping here disclosed did not amount to a search or seizure within the meaning of the Fourth Amendment.[29]

Later courts elaborated on *Olmstead*'s means requirement, of a real, physical search, by appeal to the threshold conditions defined in *Hester* v. *United States*. There the Court decreed that

> the special protection accorded by the Fourth Amendment to the people in their "persons, houses, papers, and effects," is not extended to the open fields. The distinction between the latter and the house is as old as the common law.[30]

Olmstead's means-based decision that telephone taps were not searches was taken to be the law until it appeared to be reversed, in *Katz* v. *United States*.[31] There the Court appeared to buckle under the weight of technological change, and reversed itself in a way that looked back to the values, or ends-based thinking, in *Boyd*. It argued that even within a public telephone booth could a person's right to privacy be violated, simply because by closing the door the individual feels justifiably isolated from the public world outside. And, indeed, as we shall see in what follows the courts today generally take *Katz* to have overturned *Olmstead* and to have provided a way of dealing with the issue of privacy that negotiates much more effectively a world of ever-present technological change.

Still, certain of their cases have indicated a failure to appreciate the nature of such change. In one set of cases, for example, the courts dealt with searches that uncovered small "bits" of evidence. In *Smith* v. *Maryland*,[32] in fact, was laid out what has come to be the guiding interpretation of *Katz*, albeit one which draws not from the Court's opinion itself but rather from Justice Harlan's concurring opinion. In *Katz* Harlan argued that a search has been carried out under the terms of the Fourth Amendment when the situation meets "a twofold requirement, first that a person have exhibited an actual (subjective) expectation of privacy and, second, that the expectation be one that society is prepared to recognize as 'reasonable'."[33] In *Smith* the issue was a pen register installed by the telephone company to record the telephone numbers that had been dialed from the petitioner's home. (Such a system was required before the development of modern computerized switching systems.) Smith contended that the acquisition of that list by the telephone company constituted a search. But in *Smith* the Court decreed first that "we doubt that people in general entertain any actual expectation of privacy in the numbers they dial."[34] Moreover,

> even if petitioner did harbor some subjective expectation that the phone numbers he dialed would remain private, this expectation is not "one that society is prepared to recognize as 'reasonable'." *Katz* v. *United States*, 389 U.S., at 361. This Court consistently has held that a person has no legitimate expectation of privacy in information he voluntarily turns over to third parties.[35]

In a more recent case, *United States* v. *Place*, involving the use of trained dogs to sniff for drugs, the Court concluded that the privacy interests of the people involved were not threatened because the dogs in question were capable of discovering only one thing, cocaine.[36] The Court reasoned that the violation of privacy required the collection of a larger range of information.

The issue that these and like cases raise, in the context of information systems generally and geographic information systems more specifically, centers on the question "What happens when each of these individual items of information is combined into a larger dossier?" One commentator on the issue of searches suggested the following:

> If a police officer, seeking to learn whether my car is in the garage attached to my home, lies down in my driveway and shines his flashlight through the half-inch gap between the bottom of the garage door and the garage floor, I would be annoyed somewhat by his choice of method, but my ultimate reaction would be, "So what's the big deal?"[37]

There might not be a "big deal" if this action were merely carried out occasionally by a person who kept the information in his or her head. But once each individual item of information can be incorporated into a larger, geo-coded information system the situation changes; in a fundamental sense it no longer makes sense to talk about a single-function inquiry, and the telephone numbers in *Smith* or the cocaine-sniffing dog in *Place* can only be seen as elements of a much larger system.

Autonomous technology in the courts

In this and other recent cases, it is helpful to see the court's vacillation about the relative importance of means and values and its difficulty in understanding technological change as expressive of an underlying set of assumptions about technological change and about the nature of society.[38] According to that view change is neutral and autonomous. To say that according to this view technologies are seen as neutral is to say something rather simple; it is to say that it is held that any impacts of a technology derive from its use, and not from some features inherent in the technology itself. This, well ensconced in public discourse, is the view that "Gun's don't kill people; people kill people."

The term "autonomous technology" may be less familiar, but the concept is no less prevalent; it is the view that

> Technique has become autonomous; it has fashioned an omnivorous world which obeys its own laws and which has renounced all tradition. Technique no longer rests on tradition, but rather on previous technical procedures.[39]

Hence, to believe that technologies are autonomous is to believe that they contain their own logics, their own trajectories. The development of a technology occurs in a way preordained; disk drives become larger, global positioning systems become more accurate, CPUs become more powerful, not because of human decisions, but rather because once the object comes into existence, once it is invented, those changes are built in—they are in a way part of the essence of the object.

The technical literature on the history, sociology, and geography of technology is filled with critiques of these two views; indeed, it is probably fair to say that few in those disciplines would today hold the belief that a technology can be neutral and none would believe that a technology can be autonomous. Nonetheless, the success of the courts in laying out these views suggests that they are broadly held. And more importantly, by defining the law as though technologies are autonomous, the court decisions institutionalize that way of thinking, and render it more true. Unfortunately, in the context of the conflict between means- and ends-based interpretations of privacy law, this way of thinking has several untoward consequences.

This is especially true in a set of cases directly relevant to geographic information systems because of their use of remote sensing. Here the issue is the use of technology to enhance normal means of search. In a series of cases—*Dow Chemical Co.* v. *United States*; *California* v. *Ciraolo*; *Florida* v. *Riley*; and *United States* v. *Penny-Feeney*—the courts have shown their willingness to entertain the use by law enforcement of increasingly powerful technologies, ones that call into question what some have seen as a clear boundary between curtilage and open fields, and which thereby shrink the area that can be called "private."[40]

The common-law distinction between the curtilage and open fields had in part defined those areas within which evidence could be taken at will and those in which the Fourth Amendment's requirement for warrants needed to be met. Even within the curtilage those actions which remained open to view remained public; the police, it is often averred, need not shut their eyes when looking at someone's house or yard. But if the curtilage remained subject to the gaze of the public, it was nonetheless possible to render it private, by taking the requisite steps, like planting trees or shrubs or erecting a wall or fence. In *California* v. *Ciraolo*, though, the Court decreed that changes in technology can require people to take further steps. In that case the police flew over a fenced-in backyard, at an altitude of 1,000 feet, and were able to identify marijuana plants growing in the yard. The Court held that

> That the area is within the curtilage does not itself bar all police obser-
> vation. . . . Nor does the mere fact that an individual has taken measures
> to restrict some views of his activities preclude an officer's observations
> from a public vantage point where he has a right to be and which
> renders the activities clearly visible.[41]

They continued,

> Any member of the flying public in this airspace who glanced down
> could have seen everything [i.e., marijuana plants] that these officers
> observed. On this record, we readily conclude that respondent's expec-
> tation that his garden was protected from such observation is
> unreasonable and is not an expectation that society is prepared to
> honor.[42]

In *Dow Chemical Co.* v. *United States* (rendered the same day) the Court made a similar finding, but with respect to an industrial site.

Then, in *Florida* v. *Riley* the Court went further. In this case, the police had used a helicopter, hovering at 400 feet, to observe marijuana plants through a hole in the roof of Riley's greenhouse, which was located in his backyard. As in *California* v. *Ciraolo*, the Court concluded that "Any member of the public could legally have been flying over Riley's property at the altitude of 400 feet and could have observed Riley's greenhouse."[43] And a Hawaii District Court went even further, in *Penny-Feeney*.[44] On this occasion, police in a helicopter used a FLIR (Forward-looking infrared device) to discern heat emissions from a garage, within which they believed that marijuana was being grown, aided by heat-emitting grow lights. On the basis of the photographs they obtained a search warrant.

If in *Penny-Feeney* the police went beyond the usual binoculars and flashlights, two rather different cases raise this same issue of sense enhancement in the context of a different technology. In *United States* v. *Knotts* and *United States* v. *Karo* police used beepers to identify the location of individuals.[45] The beepers allowed them to track the individuals for long periods, developing a story about where and with whom they (and their drug-producing, beeper-laden cargo) had been.

Each of these cases is slightly different, but what they have in common is the use of technologies—helicopters, airplanes, FLIRs, beepers with radio direction-finders—which belong to a technological "family." In the case of the beeper, a geographic information system would allow not just tracking but mapping of the automobiles involved. In the case of *Dow*, the cameras used *were* mapping cameras. In *Penny-Feeney* the technology used is the very technology used in satellite remote sensing. And in *Ciraolo* and *Riley* the principles used are indeed those of remote sensing. In each case the technological devices were used to enhance normal vision, to make visible what was previously not. The courts have traditionally drawn a line, not allowing the government to use technologies not readily available to the public. But here the courts seem to be accepting the view that as what is "readily available" changes, so too does the nature of the technology that the government can use.

Here, as in the cases above, of beepers and drug-sniffing dogs, the courts have argued as though technological changes are natural, and as though everyone ought to accede to them—and rethink the right to privacy. And here, too, that rethinking has generally been in the direction of diminishing the power of the individual and increasing the power of government.

But this returns us to the issue raised in *Katz*, of the nature of reasonable expectation. Appealing to the *Katz* criterion for the "reasonable expectation of privacy," the Court has interpreted "reasonable expectation" in a way that allows the reasonableness of that expectation to change along with technological change. Granted, the court has not to date allowed the introduction of satellite imagery into domestic cases (although certainly these images are used internationally, and used by the United States government in its anti-drug program in Latin America). But as the resolution of images becomes better and costs lower, these images may very well become so common that it will make sense to argue that the use of them by police is consistent with popular practice. Indeed, the rapid and widespread adoption of global positioning systems and the development of private satellite systems, now able to produce commercially available images with a resolution of one meter, suggest that this may not be as far off as many on the court have believed. Yet to say that popular practice may involve the widespread use of global positioning systems and of remote sensing is not at all to say that people will take the results of those practices to be reasonably expected.

In fact, to say that society is prepared to accept something as reasonable is at once to make a complex decision. It is to appeal to a corporate body, "society." It is to ask of that body that it envision something about the future. And it is to ask that it render a judgment about that future. Nonetheless, in a concurring and influential opinion in *Riley*, Justice O'Connor without comment redefined the *Katz* standard, of what "society is prepared to recognize as 'reasonable'."

> If the public rarely, if ever, travels overhead at such altitudes [400 feet], the observation cannot be said to be from a vantage point generally used by the public and Riley cannot be said to have "knowingly expose[d]" his greenhouse to public view. However, if the public can generally be expected to travel over residential backyards at an altitude of 400 feet, Riley cannot reasonably expect his curtilage to be free from such aerial observation.[46]

Here O'Connor has declared that "if the public can generally be expected to travel over residential backyards at an altitude of 400 feet, Riley cannot reasonably expect his curtilage to be free from such aerial observation." If the two statements seem similar, note the differences: *Katz* referred to "society" as a corporate body; in speaking of the public traveling over backyards O'Connor is surely talking at most about groups of individuals. *Katz* referred to what society is "prepared" to accept; O'Connor referred to what people in fact do. And *Katz* referred to the "reasonableness" of the expectation; O'Connor, again, referred to what people do.

In effect, O'Connor suggests that we can read off society's preferences from the current state of society and that we should see society simply as an aggregate of individuals. The first suggestion, which derives an "ought" from an "is," is simply silly; one need not read the critical literature on revealed preferences, but need only look around to find that individuals do not at all accept as reasonable a great deal of the ways in which things are. Certainly it makes little sense to deduce from the fact that one often hears helicopters overhead the conclusion that people see this as reasonable, and equally see as reasonable that what they do in their backyards or even their homes is in plain and open view. Here it is the implicit belief in autonomous technology, the belief that this technological change is occurring in a natural way, one intrinsic to it, that allows the easy conclusion that the changes are to be "reasonably expected."

This theory has allowed the courts to avoid confronting the consequences of these technological changes for the very practice of government. For example, there is in government offices today a huge store of data that are available to the public, and some of these data are now being incorporated into information systems, including geographic information systems. There is a tendency to see this incorporation as failing to raise any important issues, simply because the data were always available. But it is important to recognize that the decision to make those data freely available was made when "freely available" meant something very different from what it now means; the people who supported laws making certain kinds of data available understood the operation of those laws only in terms of the technologies available to them, and the regulations were almost all written before the development of computers and geodemographics.

In concert with the courts' implicit theory of technological change, this failure to look beyond the individual to the social and cultural value of privacy is particularly dramatic. The courts have allowed first this then that diminution of the arena within which a person can see her behavior as truly beyond surveillance, but have imagined that no structural change will grow out of the sum of those individual changes. This has led the courts to imagine that a society in which everyone's actions are under constant surveillance will be no different from one in which only some are sometimes under surveillance.[47] While explicitly embracing the ends-based *Katz* interpretation of the Fourth Amendment, the courts have in fact been basing their opinions on a view in which technological change is seen as neutral and autonomous; they have in fact deferred to a means-based view wherein means are increasingly the same.

And here, then, in a way very different from the one that we have seen in Europe, information systems and particularly geographic information technologies have been connected to a refiguration of the right to privacy. This has happened in two interlocking ways, as law enforcement has used the technologies to fight drug production and smuggling, and as the technologies themselves have seemed to embody notions of autonomous development that have allowed the use of those technologies to be seen as reasonable.

Geographic information systems, geodemographics, and the assault on privacy

In Chapter 3 I discussed the development of geodemographic systems. Combining census and other areally coded data with data about individuals and households, geodemographics creates social, cultural, and economic profiles first of areas and then of their residents. These profiles are useful for site location and political redistricting, and at the same time for drawing inferences about the households and individuals in a particular area.

These systems are, in one sense, quite unexceptionable. Their aim is simply to characterize these areas or regions of people with similar lifestyles, based on the assumption that people tend to live in close proximity with others like themselves. The characterization is, by and large, an inductive one; large amounts of data are fed into a computer, and using numerical taxonomic methods a vast number of places are grouped into a smaller number of large areas.[48] In the first-generation systems, the United States was subdivided into geographic areas of

200–300 households, each of which was characterized as belonging to one of perhaps 20–40 basic types. Newer, micro-scale geodemographic systems use the same methods, but produce areas at much smaller scales, in some cases with as few as 5–15 households. And the newest systems promise to operate at the scale of the "rooftop." In any of the systems, users claim to be able more efficiently to target their campaigns at regions of a particular type, thereby saving money, aggravation, postage, and trees.

Described in this way these systems seem quite benign. But because of the ways in which they treat information about individuals and households, they nonetheless have been a source of anxiety. This anxiety is a new version of an anxiety prevalent in the 1960s and 1970s. Then, as computers became more powerful and readily available, people began to worry that governments might, using common identifiers like social security numbers, combine disparate data files into single comprehensive dossiers on each individual. Concern about the possibility of this process of data matching led in the United States to the passage of the Privacy Act of 1974, the Computer Matching and Privacy Protection Act (which applied only to government data), and to other more specific legislation such as the Video Privacy Protection Act.[49] These acts are widely concluded to have been ineffective, largely because they allow for exceptions when matching or other data use can be classified as "normal and routine," and because what is normal and routine has been so broadly construed; as Marx and Reichman said of the 1974 Act, "Broad interpretations of 'compatible purpose' have made it possible to include nearly any government-initiated venture."[50]

Geodemographic systems do, in part, use data matching—they are exempt from the legal controls of the Computer Matching and Privacy Protection Act because they are not governmental—but, more importantly, they use a new version of data matching, called data profiling. There the key used for connecting databases is not the social security number, but rather the geographic coordinate. Geodemographic systems begin with data aggregated at an areal scale, and then typically associate those data with data about individuals or households. The association involves two steps. First, a marketer acquires information on individual purchasing habits, automobile and home ownership, voting preference, religion, and so on. Second, these data are combined with areal socioeconomic data in order to create an areal profile of residents' "lifestyles." It is here, some would suggest, that the users of these systems run up against the issue of privacy. It is not hard to imagine various situations in which a person's right to privacy may be violated, and here, to alarmists like Larson, images of Big Brother and of *1984* come all too easily to mind.[51]

Even in the absence of effective legislation it might in principle be possible to deal with some of geodemographics' potential threats. In the United States, for example, it is typical to argue that those who find their privacy rights violated can simply file a civil suit. And other countries have developed a wide range of means of protection, including data protection commissioners, ombudspersons, and the like.[52]

What many of these remedies to assaults on privacy have in common is an adherence to the fair information principles (or practices). But there are three ways in which geodemographics and geographic information systems more generally call into question the possibility of the adherence to those principles.

Undisciplined information

The first limitation on these privacy remedies is related simply to the widespread use of the systems. A quick glance at a magazine like *American Demographics* or *Direct Marketing* will show that the ones I have listed in Table 3.1 constitute only a fraction of the large and growing number of suppliers of demographic and geodemographic data in a large variety of types. The sorts of data used in geographic information systems and geodemographics have become such valuable commodities that skeptics have made a variety of attempts to regulate them while producers have pressed (often successfully) for special treatment for an industry that they regard as central to America's economic well-being. The fact that a range of organizations, from the Council of Europe to the European Community to the Organisation for Economic Cooperation and Development, have attempted to regulate the flow of data underlines the importance of these data. So too does the strength of the objections raised in 1996 as the World Intellectual Property Organization—with United States government support—attempted to establish new and sweeping rules for the protection of databases.[53]

It is easy here to imagine that something very simple is going on, and I think many casual critics have been misled by a dated and anachronistic image of computers. This image was born in the 1960s, and nurtured in the 1970s in a series of reports on the threat posed by computers to individual privacy.[54] These reports expressed widespread fears that the government would build substantial databases of dossiers on individuals and that the computer would facilitate this task by allowing the combining of separate files into larger dossiers. This fear has given rise to a guiding image, of the Benthamite Panopticon, where all is visible from a central point.[55]

But the fear of the Panopticon was based on an authoritarian image of the computer as a large, expensive, and technologically complex machine accessible to but a few. Since the early 1980s, however, the reality has been very different, because of the extraordinary proliferation of inexpensive computers. This reality was driven home just several years ago in the case of Lotus MarketPlace™. MarketPlace™ was a CD-ROM-based system that would have provided data on many of the nation's businesses (7 million) and households (120 million).

Promoted as a means of providing small and medium-sized marketers and mailers with information long available only to large corporations, the business version was introduced in the autumn of 1990, with the household version expected to follow in the winter of 1991.[56] It was not to be: As one newspaper reported:

> Lotus Development Corp. said Wednesday it will not introduce a personal computer data base that would have given businesses detailed information on 120 million Americans . . . . [O]fficials at Lotus and Equifax said they decided to withdraw it after 30,000 consumers wrote or telephoned them, demanding their names be removed.[57]

Many saw this argument as silly, aimed as it was against data that were already available.[58] But those who objected to Lotus MarketPlace™ had a point—with this system we were discarding the old view of a centralized computer in the hands of Big Brother and replacing it with something much worse. With a centralized databank it was at least possible in principle to locate each piece of data about an individual and to correct or delete those data that were incorrect. But that is not possible under the current situation; no one can hope to know who has which data about a household or individual or who has outdated or inaccurate data. If it is difficult to recall defective automobiles, it is even more difficult to recall every CD-ROM, every backup copy and every map and table that includes infringing data. In the end all that Lotus was able do for those 30,000 people who wished their names to be excluded from MarketPlace™ was to tell them that their names would be excluded from future editions. The data already in the hands of consumers were there for good. Indeed, the case of Lotus MarketPlace shows that our worry ought not to be the Benthamite Panopticon, but rather the "Mirror World" envisioned by David Gelernter, where anyone can see any part of the world, at any level of detail, at any time.[59]

The power of the visual

If this problem of undisciplined information is intrinsic to geographically-based systems, it is not exclusive to them. One can see the same difficulties with any information system, including radio and the newspaper. But a second feature of the systems, their reliance on the visual, sets them apart from other information systems, and creates a second limitation to traditional privacy remedies. Here the problem arises from the way in which visual representations are "read."

In the first instance this difference arises because those who produce geodemographics use their analyses to characterize neighborhoods or areas in terms that by and large suggest that those neighborhoods are homogeneous. The areas are typified in colloquial and highly general terms, as "Hard Scrabble" (PRIZM®) or "X-Tra Needy" and "Zero Mobility" (Niches™) or "Low Income Blues" (MicroVision®). Put in this way there may not seem to be a problem here. Still, one wonders who would not feel slighted upon finding that his or her neighborhood was titled "Zero Mobility" and not "Working Hard" or "Very Spartan."

A version of this problem arises in the analysis of statistical data where it is referred to as the ecological fallacy. The fallacy is committed when the analyst or reader assumes that the average individual within an area represents any given individual therein. Because the ecological fallacy is a way of thinking with which anyone who publishes statistical research must contend, some might argue that there is no reason to single out the visual representations produced within geodemographics.

But there is in fact a difference between the commission of the ecological fallacy in statistical work where results are displayed in, for example, tabular form, and its commission in the case of mapped data. This is because people tend to see maps as direct representations of reality in ways that tables and charts are not. There is ample evidence of this. We see it in the ways that people misread conformal projections, inaccurately concluding for example that Africa and Greenland are of similar size. Within geography, cartographic theorists have made a similar assumption; they have believed it possible to develop schemes of map representation that the average person can read directly, and without interpretation or error. As Robinson put the matter,

> Your main objective in designing a map is to evoke in the minds of viewers an environmental image appropriate to the map's purpose. . . . [So] When designing a thematic map, you must be sure that the modulations of marks and symbols you choose work together graphically to evoke the overall form of a distribution.[60]

Indeed, until very recently this view was in the United States the law of the land. It was only in the 1992 district court decision in *Mason* v. *Montgomery Maps* that the courts for the first time held that differences among maps might have an impact on the perception of the data on those maps; previously they had held that differences among maps of the same subject were merely decorative and inessential, and that any two map readers would "see the same thing" when looking at a map.[61]

Here, then, the issue of privacy is raised not simply because data are uncontrollable, or may contain factual errors about individuals. Rather, concerns arise just to the extent that it is possible to produce visual representations that any reasonable reader will directly read as associating characteristics of behavior or belief with individuals or members of households. In an important sense, such maps paint their subjects, and in colors that may be inaccurate or damaging, that may impute to those subjects behavior or beliefs which they take to be matters that should remain out of the public eye.

One might reply that the danger here is not merely with geographic information systems or geodemographics, but with maps or visual representations more generally. This misses the more critical point, that the ease of computer mapping, combined with the increasing availability of data sets, has made maps more readily available and made the possibility of privacy infringement much more likely.

Data matching and data profiles

Beyond the spread of undisciplined data and the ready appeal to the visual, geodemographic systems pose a third threat to the traditional means for the control of infringements of the right to privacy. This is through their use in the construction of data profiles. One of the earliest concerns raised by the use of computers was that of data matching. To prevent the creation of large dossiers on individuals, proponents of the Computer Matching and Privacy Protection Act of 1988 intended the Act, at least in the arena of the government, to make the merging of databases more difficult.

Yet geodemographics makes it possible to merge databases in a way that circumvents the law. These systems enable a user to create a profile of an individual not by the collation of individual data across government agencies, but rather by combining individual data with other publicly-available aggregate data, data geographically coded at the level of census block groups, postal carrier routes, and rooftop geocoding. Using a wide variety of characteristics, and using data that are publicly available, one can create a very probable image of a person.

So in an important sense the development of data profiling, where those profiles are attributed to individuals, portends an urban environment in which it is possible that any person that I telephone will, beginning with a Caller ID system (or the Automatic Number Recognition equivalent, standard and unblockable with toll-free telephone numbers) be able to move to my name, my address, and a demographic profile of me. It may well be an environment in which those in "better" neighborhoods automatically get faster customer—or even emergency—services. And it is equally possible that systems be set so that calls from people with some profiles are not answered at all.

As a result, the use of data profiling undercuts the possibility of privacy (as do the development of undisciplined information and the use of visual representations) to the extent that it renders it difficult to apply the set of fair information practices. Indeed, in a geodemographic world, where profiles are constantly being created, marketed, and recreated, it is hard to see how any individual can today know whether he or she has adequate knowledge of which data exist, has access to those data, has the ability to correct those data, or can be assured that data have been collected only where necessary.

The challenge of the virtual individual

Beyond the problems raised above, geodemographic systems are associated with a more fundamental and ongoing reconceptualization of the objects that make up the world. Within them, social and cultural groups are redefined as mere aggregations of individuals. Places are defined as locations attached to which are merely contingent sets of features or attributes. And cultures and places come to be seen as composed of or inhabited by individuals whose names and bodies come increasingly to be armatures to which are attached geodemographically-constructed identities.

But if the geodemographic form of geographic information systems represents society in an emaciated way, it represents the individual in a way that is complex and contradictory. As it creates data profiles, it constructs a world of virtual individuals or digital puppets (to use two common names for them), and these individuals raise serious problems for the maintenance of a right to privacy. Indeed, the various calls for controls on data matching were in an important sense attempts to protect the individual from government by institutionalizing that fragmentation; the fragmented individual, it was argued, was safer from government control, safer from the Panopticon.

An obvious complaint here, one that I mentioned above, is this: When decisions are made on the basis of these data profiles—themselves constructs of "real" facts about my past purchasing habits, as well as inferences drawn from aggregate data—I am being treated not like "me" but like a caricature. In the image of the 1960s and 1970s, I am being treated like a number, a category, a class. There is something to be said for this argument. After all, when the content of both news and advertising available to me is first filtered through a set of suppositions about what I "really want" as a member of some group, there is an important sense in which I have lost control of my life, in which I am no longer able to make free and informed decisions.

Recent geodemographic systems, such as Trans Union's Income Estimator (TIE), attempt to remedy at least a portion of this problem by moving to the level of the individual. According to Trans Union

> TIE does not rely on household income data. TIE does not assume that people earn the same income as their neighbors.
>
> TIE is calculated from individuals' previous and current behavior. This objective information is gathered from Trans Union's national database,

covering the spending and payment behavior of more than 160 million consumers.

TIE examines 23 behavioral characteristics identified as most predictive in establishing individual income.[62]

Many people would probably say that Trans Union's claim, that their databases are much more accurate than those of other companies, does little to assuage their fears of a loss of privacy. Most, indeed, would very likely have the opposite reaction, that this simply makes them more worried. But I would suggest that the recent move in geodemographics to more and more individualized profiles suggests a means for dealing with the privacy concerns raised by the systems.

I have suggested above that we might think of the products created by data profiling as "digital individuals." So when I apply for credit the bank officer "sees" a digital individual; when a credit card company asks Trans Union for a list of prospects it gets a list of digital individuals; and when the local pizza company sends out a mailing it sends it not to me, but to a digital individual, one who shares my name and address.

There has been a tendency to see these digital individuals as mere "puppets," and not as having any sort of reality; that is, it is commonplace to see them as images, lacking the reality that the "real me" has. But I would suggest that to see them in this way is to be seriously misled. After all, we have seen that in the twentieth century it is quite common to see individuals as having fragmented personalities. This notion was implicit in criticisms of data matching, and it is far more widely found. Erving Goffman, for example, made it a cornerstone of his work.[63] More recently, other sociologists have made related points and in the field of computers Sherry Turkle has forcefully argued that on the Internet people live nothing but fragmented lives.[64]

Indeed, I would argue that the solution to a large category of the problems raised by geodemographic systems and by certain versions of geographic information systems and the GIS family is to take digital individuals seriously. We need to see them as important, permanent features of our society and our selves. Once we begin to understand that these individuals—carrying our names, addresses, and social security numbers—are talking for us, representing us, and making decisions for us, we can see that they are very much like the fragmented parts of ourselves that we present in every part of our everyday life. The digital individual in my credit report is very much like the individual that rents a videotape or deposits a check or rides the bus. It exists in a particular place, but only for a particular purpose. Once it leaves the store or bus or bank, it for all practical purposes ceases to exist. If we take it as obvious that we have control over our actions in the video store or on the bus, we ought too to conclude that as holders of our own identity in a more continuous, physical sense we should have control over that much wider range of virtual selves to the creation of which we have been only partially willing collaborators. Here, with Foucault, we need to acknowledge that if it has been traditional to see those in the information industry as selling information about people, it makes more sense to see us as authors of our own lives, of our identities real and virtual. The information industry acts more as editors and publishers of those virtual identities.

And so, just to the extent that geodemographics and geographic information systems create the tools that render those digital individuals more real, they strengthen the case for treating them as real. And if in other ways—in their association with the belief in autonomous technology and their destruction of the principles of fair information practices—they remain a threat to privacy, in this way they offer a solution to the threat to the individual.

If this seems an odd, even eccentric, view, I suggest that it is neither. Indeed, to adopt this view is to link the issue of privacy with that of intellectual property, in the following way. As I have noted, there are two main traditions in intellectual property regulation, the Anglo-American labor-based tradition and the Hegelian-based personality theory, or theory of moral right. In the theory of moral right, a piece of property is seen as in a fundamental way an expression of the personality of the owner; indeed, one becomes fully human only by owning property, just because only in that way can one show the world who one is. This theory of property is not simply a wild fantasy. Personality-based theory was developed not as an attempt to provide a normative basis for property rights, but rather as an attempt to understand how property actually works in a modern society. Further, it is central to European systems of intellectual property; it is formalized in the major international convention on intellectual property, the Berne Convention, to which the United States is a signatory.[65] In the case of works of creativity this theory has a number of corollaries, one among them being that the work is seen as an intrinsic part of its owner and that fundamental features of the work are permanently attached to the author and cannot be alienated.

Most important, though, is a right closely related to the right of privacy, the right of divulgation: A prospective author cannot be held legally liable for failing to produce a contracted work, since to do so would be to force the author to represent herself in a way of which she did not approve.[66] So to take the digital individual as real is to offer a conception of the individual that connects the concepts of privacy and intellectual property, while appealing to well-established bodies of theory and well-established institutions. At the same time, because the theory of moral right places the individual fundamentally and intrinsically in the context of the social, it is to move away from the individualist bias that has underpinned both the theory and practice of geographic information systems and geodemographics—and that has made it so difficult to understand the problems that they raise for privacy.

Part III

LIVING WITH GEOGRAPHIC INFORMATION SYSTEMS

8

GEOGRAPHIC INFORMATION SYSTEMS AND THE PROBLEM OF ETHICAL ACTION

When we think about any organized human activity, and particularly about those activities that have substantial impacts on the natural and social world, one question sooner or later comes to mind: How ought one who is engaging in those activities to act? What things should one be certain to do, and what things ought one to avoid doing? To what system of morality ought one adhere?

When one looks at geographic information systems and more broadly at geographic information technologies, this question seems particularly daunting. And that is because it is difficult to know just where to look for guidance. If there are the obvious, general works on morality and ethics, there are also more specific ones. There are literatures on ethics and computers, ethics and planning, ethics and science, ethics and technology, ethics and business, ethics and professionalism, and ethics and government. And there are literatures, albeit smaller ones, on ethics and geography, and ethics and cartography. To which one ought one turn? And how ought one to deal with the variations in emphasis and terminology that one finds in comparing these literatures?

It seems to me that it is useful here to step back for a moment, to consider some very general features of contemporary approaches to ethical issues, and to place those features and systems within a broader context. That context is in fact familiar, because it is the very context within which geographic information systems and geographic information technologies have developed.

On the foundations of ethical activity

Most moral theories rest on a pair of assumptions. First, they assume that the actor was in some sense free. Conventionally, this means that when I do something to which ethical strictures apply, it ought to have been possible for me to have done otherwise. If a madman throws me off of a building and I scratch the paint of the BMW on which I land, most people would not hold me morally accountable.

131

And second, if I am to be held accountable for the outcome of my actions, I need to have been able to have predicted that outcome. If in turning the light switch I electrocute a person in the hotel room next door, it would be unlikely that I would be held accountable; the connection between the cause and the effect is just too tenuous.

Now, one very common way of thinking about ethics is that there are a series of ethical rules, and in a given situation I should apply the appropriate rule. A rule might be "Never lie," or "Don't intentionally hurt people," or "Treat others as you would wish that they treat you." Here a good action is one carried out in accordance with the rule; a bad action is one that contravenes that rule.

Note that in this characterization I have gone from a series of commands— "Do this—" to something very different. For to say that such and such action is bad is to make an assertion that seems very much like a factual assertion. And in fact, this is a comfortable way to speak. We speak as though events and objects in the world have a range of attributes; they have color attributes and weight attributes—and ethical attributes.

Here, though, there are two different tacks that one might take. In one the focus is on obligations, and on the individual actor as a moral agent. On this view, articulated by Kant, an action is good if I acted with the correct rule in mind.[1] By contrast, for others one measures a moral act by considering its consequences; of two acts, the better will be the one that increases the amount of happiness or pleasure in the world.[2] In the first case, simplifying greatly, a person who developed a terrain mapping system in the belief that it would allow greater access and freedom to people would be acting in a commendable way; but for the consequentialist the use of that system for guiding cruise missiles would be of overwhelming importance.

Each of these two ways of thinking is very much a matter of common sense— in truth, most of us appeal to one or another or both every day. But when we look to the ways in which these forms of discourse have been used within the realms of science and technology, we find a series of developments that seem to make the issue of the relationship between geographic information systems and ethics much more muddled.

Within the twentieth century, the ideology that has underpinned much of the discourse about science and technology has been basically empiricist. This understanding of science harked back to the work of Hume.[3] In the twentieth century, this empiricist strain appeared in the form of logical atomism, and then logical positivism and logical empiricism. At that time people like Bertrand Russell, A.J. Ayer, and C.K. Ogden and I.A. Richards began to develop a theory of meaning in language.[4] These modern-day empiricists argued that linguistic statements that are meaningful must be of one of two types. They must either be direct factual statements or be statements that are true by virtue of meaning. If they neither fall

into one of those two groups nor can be decomposed into such simple statements, linguistic utterances are just that, utterances, on par with grunts and
groans. And, they continued, because we can find no empirical content in moral
judgments, a moral judgment must, in the end, merely be an expression of
emotion. For that reason this meta-ethical view is termed "emotivism." Within the
context of logical positivism this view seemed to make good sense. It heralded a
day in which clear thoughts would be expressed clearly, and the continental and
idealist baggage, the sort of Victorian effluence, of philosophy would be super-
seded by a more modern view.

One immediate consequence of this view was to excise values from science
(and technology). It rendered them neutral; any ethical discourse was not a part
of science, to the extent that it shared none of the methods of science. Real,
meaningful language, to use a favored image—and one that we have seen in
Chapter 2—maps onto the world, a visible world. Good and bad are nowhere to
be seen. Or at least, they cannot be seen *as* good and bad. Rather, if we are to
judge the desirability of a particular action, on this view we need to look at those
consequences in terms of some visible measure.

And indeed, this is very much the model that underlies more current
economic discourse; there, value is translated into money, into income or profit,
and to maximize one or the other is on a neo-Benthamite calculus to do good.
Further, it is the model that underlay geography in the era of the quantitative
revolution; the classics of that era, by people like Bunge and Harvey, all speak in
the language of empiricism, and emotivism.[5] And so, too, did it underpin the
ameliorist planning mentality within which geographic information systems first
developed; there, the intent was to use geographic information as a way to better,
more efficient use of resources, and thereby to remove slums and ghettoes,
improve transportation and education, and generally improve the social welfare. If
those developing the systems harbored hidden views that there really is a good
and bad—and I am sure that they did—their work was very much in accord with
the empiricist/emotivist mainstream of science. The tools, methods, and everyday
activities were seen as strictly neutral, as neither moral nor immoral, but non-
moral.

Yet the striking thing about the everyday practice of science, or of using
geographic information technologies, is just the extent to which those who do it
see themselves as operating quite firmly in accordance with a set of core values.
This, after all, was just the argument that Merton made. At the same time, I
have suggested that the use of geographic information systems is a project that
undercuts each and every one of those values. If this is the case, what is to be
said of the nature of ethical action in an era of pervasive geographic information
technologies?

Ethical action in an era of pervasive geographic information technologies

I would argue that in the face of these technologies this simple model of ethics—and it *is* a simple model—fails us. (Indeed, it has always failed us, just because it was based more on an idealized image than on an actual inquiry into the nature of moral behavior.) In what follows I shall delineate four areas in which it fails. It fails because in many of the cases described by geographic information systems prediction is impossible; and it fails because it involves an inadequate understanding of the nature of human thinking, of the geographical world, and of the user of geographical tools. In each of these areas I refer back to previous chapters.

Unpredictable systems

The user of a tool cannot, according to the conventional view, be held responsible for the consequences of using that tool if those consequences could not have been predicted. Consider a geographic information system as a combination of software, hardware, and data. Early GIS programs were often small indeed; in today's world of program sizes in the dozens or hundreds of megabytes, it may be difficult to imagine but some programs could be fit on 100 or fewer 80-column cards. In those cases, where the system involved a limited number of operations on data of a limited number of types, a GIS was not very different from a statistical table; one could scan the data to find problems. Further, problems could often be easily fixed, because the person using the program was often the person who had designed and written and compiled it. And if undertaking that repair was often a nuisance, it was typically a project that could be completed in short order.

Today the situation is very different. Put simply, software is very much larger than it was. But the difference in size also means a number of other differences. The software was almost undoubtedly written by a number of people. In a program as complex as modern GIS packages there is almost certainly no one who has read every line of code, or even who *could* read every line. Indeed, the program was probably written in a number of places, perhaps thousands of miles apart, and may have been written by people who speak different languages. And it may well have been written over a long period of time; in what seems a turnabout from the usual pattern of expectation, "soft" ware has a much greater life expectancy than "hard" ware. The notorious "Year 2000" problem, where current software lacks the ability to use dates after 31 December 1999, is a clear case in point.

But even for those who write smaller programs, the development of a division of labor within the programming industry has meant that the effect is the same. The

134

development of object-oriented programming has meant that even programmers can in part of the process be shielded from the code itself; they may never need to see it, and because of trade secrecy restrictions may *not* in principle be able to see it.

When we turn to software purchased or leased by a person or organization that is not the end-user, from an ethical standpoint things become even more difficult. The user is almost always confronted with a package that is in effect a black box. Not simply the software, but the statistical and representational algorithms written into the software are hidden.

Indeed, and as we have seen with the issue of ownership rights, the complexity of modern systems—where they may be written by individuals or teams; owned by various forms of legal entity, including individuals, partnerships, and corporations; leased, sold outright, or offered as part of a service—can make it almost impossible to make sense of who has responsibilities for a system. If in the case of liabilities the courts tend actually to allocate percentages of blame to different responsible parties, this seems a mockery of the commonplace sense of morality. Most teachers, after all, would gasp if a student claimed to be only 17 per cent responsible for a paper's having been late.

If this is a problem with software, it is also a problem in the case of data. In a way, the image that most of us have in mind when it comes to data is one that comes from the laboratory. As a scientist I devise a hypothesis, then a means for testing it, and finally the hardware for doing so. In my laboratory, under controlled conditions, I run some experiments, and from them collect data. I am then able to submit them to analysis.

In geography this is an ideal that few of us can achieve, but it remains as a model, where there is at work a kind of distance-decay function; the farther we are from the source of the data, the more suspicious we are. At the same time, the more we can account for the steps in collecting the data—and the more we see those steps as formalized and perhaps mechanized—the more comfortable we are with those data.

Yet as we move to larger and larger data sets, and as we extend the reach of analysis back in time as well as across space, the more questions arise about data quality, just as the more it costs to guarantee that quality. The problem arises in one way when the data have been collected by humans, in weather stations or in censuses. There we need to deal with the vagaries of human nature and cultural difference, with the desire of people to look as though they are doing their job when they are patently not, and with the desire to mask failure and highlight success. Granted, one can introduce standards into the process, as those involved in spatial data have done. But in a sense, this merely moves the problem back a step. For now the question is how do we deal with these same vagaries in the case of those applying the standards? And if we set up a meta-standards organization, the question is how to judge the standards judges . . .

We may be inclined to believe that automating the data collection process will resolve the problem. The data will come streaming in, untouched by human hands. But of course, it does not. In the end, the data come from machines that have been calibrated by the same sorts of people who might have collected them. Moreover, the machines used to collect the data are themselves constructed by humans, in the same sorts of settings, and with the same sorts of complexities, that attend the creation of software. Here, too, some have seen the adoption of standards as a solution; and indeed, we increasingly see signs on buildings and trucks touting compliance with ISO 9000. But here again, the same problem of regression applies.

Finally, much the same is true of hardware. If we tend to imagine computer hardware on the model of the adding machine, where you put in the data, turn the crank, and out come the results, the reality is rather less comforting. Particularly where one is dealing with very large numbers or with complex operations, the algorithms built into the hardware for dealing with computations render the hardware itself much more fluid than the usual image would suggest. We were reminded of this a few years ago when the Intel Corporation was forced to admit a problem in the way its Pentium chip handled certain floating-point operations.

So in effect, the size and complexity of software, data, and hardware mean that even if all were under the control of the user of the system, that user would not be and could not be justified in claiming to understand them well enough to be responsible for them.

The human as computer

Does this mean that the possibility of morally acceptable behavior by those using geographic information systems does not exist? No, but in order to see why we need to rejoin an issue dealt with in Chapter 2. There I pointed to the common appeal in computer systems to interrelated conceptions of language and rules. On this view, the elements of language map onto or mirror the world. When I say "cat" I have in mind a concept of "cat" that matches the English word "cat," and that in turn refers to a class of objects in the world. Words need not refer to classes of objects; they may also—depending on the flavor of this view that you prefer—point to individuals, via names, or events. Still, the point is the same.

On this view mental concepts, language, and the world itself share this; they are organized in terms of rules and relationships expressible in rules. So, for example, one rule states that it is impossible, as a conceptual matter, for an object to be both all red and all blue. Another rule states that in a proper sentence in English, the verb and subject agree with respect to number. And a third rule states that if an object is at rest, it will remain at rest, in the absence of some outside force.

Two points here are extremely important for the issue of ethics. The first concerns the way that this type of thinking rests on a series of spatial metaphors. We imagine, after Descartes, the mind as a mental space, separate from the physical space of the world.[6] And we imagine mapping concepts and words onto the world. In each case the relationships can be seen as universal, as having the character of logical and mathematical connectives. Leibniz, in fact, believed that a statement was necessarily true if it was true in all possible worlds, necessarily false if false in all possible worlds, and contingently true if true in some. As a result of this appeal to rules, this way of thinking lent itself to an easy translation, where the mind came to be seen as a computer.

Second, this way of thinking about the mind as operating in terms of rules led, as far back as Aristotle, to the view that just as there are the rules of formal logic which deal with what is, there must also be a set of rules for dealing with ethical judgments.[7] On that view, one could think of an ethical judgment in the following way:

If you want to be a good person you must be generous;

If you want to be generous, you should donate a portion of your income to the poor; therefore,

If you wish to be a good person, you ought to donate a portion of your income to the poor.

This formulation has the advantage of making ethical judgments seem very much like judgments about migration patterns and the like. Here, in our everyday life we are imagined, scientist- and computer-like, to be engaged in a constant process of planning and scheming, devising strategies—if we are to act morally—for adhering to the right principles and achieving the right goals.

But there is a problem here, and one that brings us back full circle. The problem is that although when we look at this model of the thought process, and then at the traditional model of computing, it appears a natural move from the first to the second, this is not what in fact happened. Rather, as I noted in Chapter 5, Philip Agre has shown that the modern model of computing was itself derived from a model of work and the division of labor, between the managing mind and the laboring body.[8] If, indeed, we follow this line of thought and take the popular acceptance of the mind–body split (as opposed, say, to the soul–body) as a recent phenomenon, and one based within industrial society, no wonder we have trouble using that model to understand ethical issues that arise as that industrial model is recast through the development of the computer and of information technology.

But this brings up a related issue, and one that turns out to be critical to an understanding of the nature of ethical action. For not only is it the case that the popular acceptance of the distinction between the mind and the body followed the development of modern systems of management. So too did the acceptance of the centrality of rules to ethical discourse follow on the invention of writing, some two thousand years before, and the spread of literacy after the invention of the printing press.[9] Prior to that time, the primary repositories of guides to ethical behavior were in the forms of narratives, of the sorts of stories that we see as far back as Hesiod's *Works and Days*.[10]

And indeed, it is now becoming clear that just as for many years the teaching of language operated—badly—in terms of a model of linguistic object and rule, so too has the nature of ethics been misconstrued when it has been seen as a subject that can be captured in terms of rules. Rather, much about ethics, and about learning ethics, is a matter of learning the appropriate stories, brief or not-so-brief accounts of what happened when someone did something. In the end, the appeal to rules that Piaget saw as a final stage in maturation was simply one that goes along with particular parts of a particular society.[11] In the time of Hesiod, as in much of everyday life today and in every age, the written is far less important than the oral, the rule than the story.

The place of place in ethics

If this seems a wild assertion, it turns out to have been a central conclusion drawn in a wide range of studies of the history of science.[12] Indeed, if the popular view of Thomas Kuhn's theory of paradigms sees those paradigms as sets of ideas, a closer reading will show that for Kuhn a paradigm consisted in large part of an interconnected bundle of stories and practices.[13] And in an area perhaps closer to the work done in geographic information systems, Donald MacKenzie has shown the ways in which stories—about how close is close—were connected with the development of standards of accuracy in guided missile systems.[14]

Turning again to the issue of rules, I noted in Chapter 3 that Wittgenstein asserted the following:

> 'How am I to obey a rule?' . . . If I have exhausted the justifications I
> have reached bedrock, and my spade is turned. Then I am inclined to say:
> 'This is simply what I do.'[15]

At the same time, he was at pains to show that there can be no such thing as a private rule; all rules—and all language—must, in the end, be public. So the answer "This is simply what I do" typically refers to "This is simply what we do," and this is as true of stories as of rules. Put another way, everything we say and do makes sense because it is situated; to say the wrong thing at the wrong time is to risk misunderstanding, or worse.

But I should like to point here to a particular class of statements of the form "This is simply what we do." These are statements of the form "This is simply what we do *here*." And just by virtue of the "here," those statements refer in an irreducible way to places. They refer to what people take to be acceptable and unacceptable, right and wrong—here, in a place.

They do something else; they refer to a way in which normative statements, statements about what is good and bad, right and wrong, beautiful and ugly, are connected to places. These statements of the form "This is simply what we do here" do something more; they point to the way in which places, at once the subject matter of geographers and the sites of production and use of geographic information systems, are themselves constituted by these normative actions. Indeed, *any* description of a place that is a description of a place must in the end incorporate descriptions of people's actions in that place. It must thereby display the normative character of the place and those actions.

Note, though, that from the point of view of a person interested in understanding the ways in which geographic information systems raise ethical issues, the normativity of places is important in two ways. To say that in giving an account of a place we need to give an account of the actions and practices that occur within that place is to say something about what we study. But it is also to say something about the way in which we do so.

And in fact, we see just this in accounts of the work of scientists, as in works by Latour, Shapin and Schaffer, and Traweek; in each case there is a strong sense that in studying scientists one needs, as Latour put it, to "follow them around," and that one does this in particular places.[16] Nor do we need to go outside of geography to see this; we need only look at—or listen to—the various accounts of life at the Harvard laboratory, at Washington, at ESRI, at the Defense Mapping Agency or at MLMIS.

In a provocative account of the nature of ethics and ethical discourse, Alasdair MacIntyre turned his attention to just this emplacedness of ethical discourse.[17] Describing the very different society of classical Greece, where to be ethical was to act in accordance with and from a sense of virtue, he points to the ways in which there, as elsewhere, the existence of ethics presupposes the existence of a community within which that discourse can be said to make sense.

If for Aristotle this community was well-defined—it was spatially tight-knit, and excluded various people, including women, slaves, and barbarians—today such communities continue to exist, albeit in different forms. In any account of graduate education or of laboratory life one must be struck by the ways in which what is acceptable and what is not are taught, and by the ways in which what is acceptable is nuanced, where a first year graduate student, a full professor, and a technician play by very different rules, or as I would prefer, are told very different stories.

Living with geographic information systems

But what separates us from Aristotle is just the fact that we have allegiance to a welter of such communities, to department and office, family and neighborhood, state and country. And indeed, it is just this that provides both a solution and a problem for those who wish to take seriously the ethical issues raised by the systems.

As I noted earlier, a central difficulty in applying traditional notions, where we apply rules that arise from a sense of duty or from a belief in the desirability of a set of consequences, is that the attempt to apply rules seems to involve us in conflict and regress, while the appeal to consequences seems to paralyze us. The systems are too big and complex, their workings and outcomes unfathomable.

Yet when we look at the everyday activities of a person who is using the systems, we typically find a well worked out set of approaches to the problems that those systems raise. Very often, when faced with a choice, one will advert to some standard of "acceptable practice." And this makes perfect sense, just because in learning a job we by and large learn not "what the rules are" but "what to do." Similarly, when scientists are challenged in court, they appeal to "standard practice" or "acceptable practice."

So in fact, in the use of geographic information systems there exists a substantial ethic, in the form of the lore about who did what, and who ought not have done what. Yet the question remains, when faced with such a system, what ought one to do? More particularly, how should one respond when faced with issues like:

Should I use computer software if I do not understand the ways in which it works?

Should I teach students to use geographic information systems if they appear not to have an understanding of basic geographic principles?

Should I develop a system if I know that it will be able to be used by people who appear not to have an understanding of basic geographic principles?

Should I take donations of computer software, and use that software in teaching students?

Should I use data that I have not collected or verified myself?

Should I work on a system that will track the everyday activities of law-abiding citizens, including those not convicted of any crime?

Should I develop a system if I know that it may be used to support scientific research likely to support positions about global warming or animal extinction with which I disagree?

What if I know it will be misused?

Should I work on a system of terrain modeling that I suspect may be used for guidance in missile systems, or for designing military plans to undercut regimes unpopular with my own government's?

What if it will be used in a regime that practices widespread torture and infanticide?

What if my software is likely to fall into the hands of that regime?

Should I support the development of systems for political redistricting, if I know that they will be used primarily for a process of gerrymandering?

And what if I know that that gerrymandering will be used by racist politicians to prevent African Americans or Asians or Latinos from having a voice in the political process?

Should I work on geodemographic systems that will be used to improve the accuracy of "rooftop geocoding," and further to manipulate consumer behavior?

This is not, of course, an idle set of questions; and in fact, these are just the questions that I have in various ways faced, or seen others face, in the last few years.

I have already laid out one portion of a way to deal with these questions. We all deal in our work with issues of right and wrong, good and bad. Those of us who manage to hold on to our jobs typically have a sense of what is acceptable and unacceptable in the workplace. But for many people that is the end of the story. Indeed, in some work places, as at the National Intelligence Mapping Agency (NIMA, formerly the Defense Mapping Agency, or DMA), the division of labor, where each person works only on a very small portion of a project, never seeing the entire project or knowing its aims, has been a means not just of maintaining security, but also of preventing the development of an ethical discourse even at the scale of the workplace.

Still, many people leave their work at work, and imagine that having done their jobs "in accordance with standard practice," they have done all that they need to do. This is of course a sad caricature of Aristotle's communitarianism. So one solution is for those who create and use the systems to attend more carefully to the other communities to which they belong, and to ask themselves about the relationship between the products of their work and those communities.

Consider the following example, which I take from Richard Sclove.[18] Imagine that I begin increasingly to shop via the Internet. What are the consequences? Trivially, I save gasoline and time. I may have a larger selection, and find better prices. But beyond that, there are other likely consequences. If others do the same, what will happen? Local businesses will go out of business. But businesses do not simply function to sell. They are meeting places, places for gossip and political talk. They are places for community building and place building. In fact, in just *going* shopping I am engaging in those processes, and as Jane Jacobs famously pointed out, making my neighborhood and city safer and more livable.

This, of course, is a single example. But it gets to the heart of the problem for those who create and use geographic information systems. And that is because the systems can easily come to replace other systems, systems that operate within the physical communities in which we all live. To take an example more directly related to geographic information systems, through the 1960s credit reports were handled through local agencies, and almost every town had one. When a person applied for a loan, the bank called the agency, and the agency sent someone out to snoop, to ask questions of neighbors and employers. In the last twenty years, though, those agencies have been replaced by a very small number—three, really—of very large credit reporting agencies, which have increasingly automated the credit reporting process. As I suggested in my discussion of profiling, there are a number of consequences to this change, but certainly one is to recast the ways in which people think about neighborhoods and communities, to recast their connections with the institutions in those places.

Finally, it has been widely reported that in the talks in Dayton about Bosnia, the United States won the day by presenting an animated, virtual fly-over. It was widely reported that the reaction of those present was that after viewing it they "Saw that a one-mile corridor wouldn't be wide enough." But there is another side to the story; surely those on the other side must have been stunned to see the detailed knowledge that the United States had. Many of us have had the experience of flying in an airplane over our homes, and the thrill of pointing out "That's where I live." Few of us, though, have been in a position of having a hostile military agency do virtually the same thing. If the first response to such a viewing is "Gee whiz," a more sober one must recognize that the power of surveillance *is* power, and that those at the other end of the viewer cannot help but feel beleaguered.

Yet here, it seems to me, we need to face up to our inability to remain conscious of very many of the communities to which we belong. Work, family, neighborhood, state; at some point we reach the impasse faced by consequentialists, the matter seems just too complex. And here we need more soberly to consider real alternatives to the geographic information systems with which we are familiar.

9

BEYOND PaleoGIS?

Over the past several years, an increasingly large series of critiques and even attacks have been launched at geographic information systems. Some of these have appeared in a special issue of *Cartography and Geographic Information Systems*, which collected essays presented at a conference at Friday Harbor Washington; others appeared in the more ambitious and empirical *Ground Truth*.[1] More critical were individual pieces by Neil Smith and Bob Lake;[2] indeed, many of the wider range of students of the systems found that virtually every encounter with a new GIS practitioner turned, sooner or later, to the issue of how the interviewer felt about Smith's piece, which many felt to be a hatchet job.

In the years since Smith's work, the literature on the systems has come to be more balanced, or perhaps more forgiving. Nonetheless, as the reader of this volume has seen, there has remained in the literature a strongly critical element. But this criticism has increasingly come to involve not Luddite calls for an abolition of the systems, but rather calls for a new GIS, for what some have called a GIS_2. Connected in part to another goal, of developing systems that might be more appropriate to public participation, by neighborhood and environmental groups for example, advocates of GIS_2 have suggested the possibility of a reformulation that will meet the needs of these groups, while at the same time skirting the various problems that they see as attending current, or what might be termed "PaleoGIS."

Although the calls for a GIS_2 have been few, and muted, it strikes me that they represent an interesting attempt at rethinking geographical explanation, and the relationship between geographic information technologies and society. In what follows I shall first lay out the central goals enunciated by advocates of GIS_2; as we shall see, their critique in some important ways parallels the analysis that I have offered here. Second, I shall suggest some of the difficulties that would attend attempts to attain them. And finally I shall suggest some ways of moving beyond the PaleoGIS–GIS_2 conflict.

On GIS$_2$

More complete representation   Central to GIS$_2$ has been the notion that within PaleoGIS there is a systematic underrepresentation of various features of the world. Just as historical accounts of earlier periods systematically underrepresent people who did not own property or were illiterate (this often included women), geographic information systems underrepresent those who do not have a location. Transients, the homeless, undocumented workers, and perhaps gypsies, all come to be systematically underrepresented. If this seems a quibble—some counter that to be missing in this way just means that one gets less junk mail—it in fact may have serious consequences. In the past, those who have not had a real attachment to places have been branded as uncivilized, as less than human; and this has been just the criticism leveled at Jews.[3]

At the same time, certain sorts of places are underrepresented within PaleoGIS. Developed within a modernist spatial order, the systems exhibit a preference for places that can be represented with the point, the line and the bounded area. But if, as I suggested in Chapter 8, we begin to see places as loci of human actions, it seems less than clear to advocates of a GIS$_2$ that something like Figure 9.1 would get at that understanding of places.

If these are criticisms of the representational capacity of particular parts of the world, critics also offer more general criticisms. They see the systems as unable

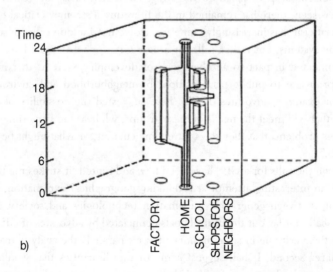

Figure 9.1 Place

Source: Allan Pred (1981: 11).

easily to represent the normativity of cultures, the ways in which cultures are defined by conceptions of right and wrong, good and bad, accurate and inaccurate. Nor, they argue, can one use the systems to represent people, to show "who they are." At best, we are left with representations of the average, the mean, with profiles and caricatures.

If advocates of GIS$_2$ see PaleoGIS as unable adequately to represent people and places, neither do they believe that the systems adequately represent time and nature. Referring to the sorts of arguments that I offered in Chapter 2, they argue that basic elements of the human experience of time, including the narrative order expressed in oral histories, lose their connection with the world when incorporated into something as cold and mechanical as a geographic information system. And similarly, the sorts of connection that a person like Aldo Leopold saw as existing in the natural world cannot be easily represented within the confines of a PaleoGIS.[4]

And finally, PaleoGIS seem unable easily to represent the place of the person who created the system; we seem, in the end, to have that ambivalent view from nowhere/view from a disembodied point. This inability, from the point of view of advocates of GIS$_2$, has the unfortunate consequence of enhancing the prestige of the viewer, just by making her (or presumably, him) seem a neutral observer.

More accurate representation   Related to the desire for more complete representation is a desire for more accurate representation. In a sense, of course, the last point, about a desire for more complete representation of the creator of the system, is also a call for more accurate representation (Figure 9.2). Here the desire is not just that the nature of the creator of a system be visible, but also that that person's—or corporation's—interests be made more explicit.

But there are other aspects to this concern. One finds a desire for systems that are more open with respect to the ways in which they work; this is expressed in a desire that the inner workings of software should not be hidden under the shield of trade secret law or employee contracts. At the same time, it is expressive of a desire that the systems be not only open but comprehensible; indeed, the average geographer would be as likely to be able to make sense of a telephone call from an avocado tree as of the source code of one of today's mammoth GIS packages.

And the desire is that not merely in general, but also in particular cases, the workings of the systems, including the ways in which statistical models were implemented and the ways in which data categories were reached, should be made more open. All of us who analyze and represent data know that today, using computer systems, it is far easier than it was using manual systems to manipulate data categories and representational schemes, and to work backwards from a desired representational outcome to data categorization; the desire here is to forestall the temptation to engage in that process.

HOME HOPPERS

Figure 9.2 Niches
Source: Polk Direct.

Accessibility In the case of accessibility, too, there is an overlap. Here, if the desire is for accessibility to information about the ways in which systems work and to the availability of systems less encumbered by intellectual property restrictions, it is also for broader accessibility to the systems by more groups. This concern, in part already fulfilled in some arenas, is expressed in the desire for empowerment of people in political redistricting disputes with the dominant political parties, in environmental disputes with corporate and government interests, and in urban-neighborhood disputes with a wide range of larger and often outside interests.

If we see a certain amount of success in each of these cases, the desire here is for more success, and for the routine availability of workable systems. It is also for the availability of systems that will enable political groups to deal on an even

146

footing with other larger interests; where a government can offer three-dimensional, real-time terrain representations and a local political group is offering black-and-white maps, it is clear who has the rhetorical advantage.

Power And this desire for power is expressed more broadly in GIS_2. Often underlying the first three principles is a desire to diminish the increasing porosity of the boundary between industry and the university, by controlling the introduction of closed software and non-standard hardware systems into the teaching environment.

Similarly, one often finds a desire to dissolve the ties between the military and the university, and to separate geographic information systems from their traditional connection with military development. In both areas, there is a desire to increase the autonomy of actors in the academic community, and to gain, or regain, a sense of lost independence.

Privacy A fourth area of concern is privacy. At the forefront here is the desire to undercut the increasing ability of business and government, using geocoding, to interconnect previously separate data systems. This, of course, is the process used in geodemographic systems.

Closely connected is the desire to control the proliferation of systems of data capture, as in the systems widely used in supermarkets, service stations, ATMs and wireless telephone systems, where it is possible to track the user of a particular credit card, automobile, or other technological device.

And finally, the desire is to diminish the proliferation of systems of digital individuals that have been constructed through the interconnection of spatial systems and the input of information obtained through data capture systems.

Indeed, one might well see this, along with the ideal of public participation, as right at the heart of the desire for a GIS_2—and the two can be seen as interconnected: with the call for more public participation, a desire for positive action; and with the concern with privacy, a desire for control over the actions of other, often corporate, actors.

The creation of community An overriding theme in discussions of GIS_2 has been the possibility of what some would characterize as a utopian or nostalgic return to a premodern past. At the heart of much of this discussion has been the hope to undercut the drive to individualism that seems to underlie so much of everyday life, and at the same time, to reject the false communities constructed in geodemographic systems, in redistricting schemes, and the like. One finds a desire for a community based less on shared patterns of consumption than on shared connections to place, neighborhood, nature, and the like.

And here one sees, too, a rejection of the ameliorative politics that have been at the heart of PaleoGIS, a desire to start anew and fashion systems that will help us salvage that which exists in place and nature, and that will use the power of computers to fashion new tools for the maintenance and improvement of everyday life.

Beyond the mystique

This desire for a new geographic information system is expressed in various ways. We see it in part in the texts that have appeared about PaleoGIS; we also see it in the development within universities of projects to promote public participation GIS. My characterization above has, of course, itself been a bit of a fiction; it is unlikely that anyone holds each of those positions, though it does strike me from interviews, conferences, and readings that each position has strong and vocal adherents.

However much I sympathize with certain of the critiques and desires expressed in this characterization, I nonetheless think that one needs to see that the entire GIS_2 project is in part based on the granting to existing geographic information systems of a kind of mystique. This has damaging consequences, and in two ways.

First, I would suggest that at least with respect to some of the concerns that I have mentioned, the desire to construct a new GIS, a GIS_2, is based on a misapprehension. It is in fact based on a failure to see that such systems already exist. They exist in computer bulletin boards and chat rooms. They exist in travel guides and classified ads. They exist in bulletin boards at supermarkets and delicatessens. They exist in neighborhood newspapers. Indeed, they exist in the enormous store of knowledge in any community.

If this seems an extravagant statement, consider what they are like. They share a great many of the features desired by advocates of GIS_2. They are local. It is difficult—this is probably an understatement—to incorporate them into larger systems of surveillance. They are inexpensive. They are open. They are controllable. They contain useful information.

If all of this is true, why have the advocates of GIS_2 not seen it? They have not seen it because they have been taken in by the mystique of what they in some ways think of as their archenemy, PaleoGIS. They have insisted on modeling an alternative system after the very one that they are rejecting. In doing so they have made their own task all the more daunting. But they have also devalued and undercut the very sorts of things that they say they want.

I don't mean to suggest that the advocates of GIS_2 are a group of dimwitted fools. Rather, I merely suggest that their failure to see the existence of ready alternatives points to the power and authority of that mass of technologies and images that we term "modernism." Together they determine, even overdetermine, our decisions to look this way and not that, to see this and not that.

There is a second way in which this mystique is involved in thinking about current geographic information systems and their alternatives. If we ask people in almost any university geography department what they believe to be the most important technological development within geography in recent memory, most will almost certainly respond that it has been the development of geographic information systems and geographic information technologies. They will point to new labs, to discarded pens, sheets of rub-on lettering curling in a corner, the occasional T-square. And they will point to rows of workstations, digitizing tablets, plotters, and the like.

And yet, the fact is that geographic information systems have had very little impact on the practice of geography. Consider a simple fact: In 1996, of 7,271 members of the Association of American Geographers, 1,181 claimed to be experts in the use of the systems. Several people have mentioned this statistic to me, responding in awe to the large number claiming to be experts. Yet looked at from another perspective, about 6,000 did *not* claim to be experts. But note, too, that of the several thousand people who each year present papers at the Association's annual meeting, 100 per cent present abstracts prepared on a computer.

In fact, if one wants to see the most important technological changes affecting the practice of geography, they certainly are: Word processing, the Internet, electronic mail and the fax, preceded by the jet airplane, the television, and the telephone. One may argue about the details, but certainly, with the exception of the fax, each of those developments dramatically changed the practice of geography for virtually everyone. Remarkably, these developments are virtually invisible, unheralded.

They are, of course, unheralded for a number of reasons. They seem ubiquitous, prosaic; everyone has a telephone. No one donates airplanes to academic departments. And most obviously, geographic information systems have seemed to be quintessentially geographical.

Yet as we have seen, there is a negative side to the introduction of the systems into the academy; there is the resulting commercialization, the imposition of outside standards for data collection, a loss of authority, the loss of a sense of independence.

But I would end by pointing to a much more fundamentally negative feature of the use of geographic information systems within the university. Just to the extent that geographers have focused on *geographic* information systems, they have failed to attend to the geographic implications of *information* systems. The time–space convergence described by Don Janelle in 1969 and echoed in David Harvey twenty years later is upon us.[5] As a wide range of scholars have shown, the development of the systems is associated with a transformation of communities, work, and leisure, just as it is associated with a rethinking of every aspect of life, where all that is noticed becomes information, information becomes intellectual property, and there is no residue. In these ways, information systems need to be seen as fundamentally transformative. But as these changes have been occurring around—and within—the academy, geographers have worried over their terminals, mapping the last millennium.

NOTES

INTRODUCTION

1 Donald Janelle, "Spatial reorganization: A model and a concept" (1969).

1 REASON AND LANGUAGE IN GEOGRAPHIC INFORMATION SYSTEMS

1 National Institute of Standards and Technology, *Spatial Data Transfer Standard* (1992).
2 See Rudolf Carnap, *The logical structure of the world: Pseudoproblems in philosophy* (1967); Gottlob Frege, *On sense and reference* (1952); and Bertrand Russell, *Descriptions* (1920).
3 I have in mind the series of revolutions beginning with Thomas S. Kuhn, *The structure of scientific revolutions* (1970) and Norwood Russell Hanson, *Patterns of discovery* (1958), and then extending in a range of directions, especially to the sociology of scientific knowledge and the rhetoric of science.
4 Ludwig Wittgenstein, *Tractatus logico-philosophicus* (1961). Wittgenstein was not alone in this; related views can be found in Gottlob Frege, *On sense and reference* (1952) and Bertrand Russell, *Descriptions* (1920) and "On denoting" (1956).
5 Norman Malcolm, *Ludwig Wittgenstein: A memoir* (1966).
6 Plato, *Theaetetus (1959)*, *The Sophist* (1935); and Plato, *The Republic* (1945).
7 St. Augustine, *Confessions* (1961:29).
8 René Descartes, "Discourse on the method" (1971) and "Meditations on first philosophy" (1971); and John Locke, *An essay concerning human understanding*, (1975). Among the many other such works, see also those on artificial languages, such as Antoine Arnauld and Claude Lancelot, *The Port-Royal grammar: General and rational grammar* (1975) and John Wilkins, *An essay towards a real character, and a philosophical language* (1668). For a discussion of language and modernism more generally, see Timothy J. Reiss, *The discourse of modernism* (1982); for a discussion of the attempt to develop artificial languages see Hans Aarsleff, *From Locke to Saussure: Essays on the study of language and intellectual history* (1982) and Mary Slaughter, *Universal languages and scientific taxonomy in the seventeenth century* (1982). I discuss this issue at more length in Michael R. Curry, *The work in the world: Geographical practice and the written word* (1996).
9 Alfred North Whitehead and Bertrand Russell, *Principia mathematica* (1925–27).

151

10 Noam Chomsky, *Rules and representations* (1980); and Claude Levi-Strauss, *The savage mind* (1968).

11 Mark Blades, "Navigation and wayfinding in information systems" (1993); Myke Gluck, "Making sense of human wayfinding: Review of cognitive and linguistic knowledge for personal navigation with a new research direction" (1991); M. White, "Car navigation systems" (1991); Arif Merchant, "Expert system: A design methodology" (1992); D.A. Waterman, *A guide to expert systems* (1986); T.J. Kim, L.L. Wiggins, and J.R. Wright, *Expert systems: Applications in urban planning* (1990).

12 A.M. Turing, "Computing machinery and intelligence" (1950).

13 Norman Malcolm, *Ludwig Wittgenstein: A memoir* (1966: 9).

14 Ludwig Wittgenstein, *Philosophical investigations* (1968: Part I, §199).

15 This argument was laid out in Ludwig Wittgenstein, *Philosophical investigations* (1968), and is elaborated in more detail in his *Remarks on the foundations of mathematics* (1983). In David Bloor, *Wittgenstein: A social theory of knowledge* (1983), there is an attempt to spell out more explicitly Wittgenstein's often difficult analysis, but with mixed results. From rather a different perspective recent work on the rhetoric of mathematics supports this same point; see, for example, Philip J. Davis and Reuben Hersh, "Rhetoric and mathematics" (1987).

16 Gilbert Ryle, *The concept of mind* (1949); Michael Polanyi, *The tacit dimension* (1983).

17 Lee Leung, "A prospectus based on fuzzy logic and knowledge-based geographical information systems" (1990); Daniel Z. Sui, "A fuzzy GIS modeling approach for urban land evaluation" (1992); G.B. Hall, F. Wang, and Subaryono, "Comparison of Boolean and fuzzy classification methods in land suitability analysis by using geographical information systems" (1992).

18 Alan Garnham, *Artificial intelligence: An introduction* (1987).

19 David M. Mark, "Representation of geographic space in natural language, minds, culture, and computers" (1990).

20 George Lakoff, *Women, fire, and dangerous things: What categories reveal about the mind* (1987); George Lakoff and Mark Johnson, *Metaphors we live by* (1980).

21 A. Newell, J.C. Shaw, and H.A. Simon, "Empirical explorations with the logic theory machine: A case study in heuristics" (1957).

22 Judea Pearl, *Heuristics: Intelligent strategies for computer problem solving* (1984).

23 Ludwig Wittgenstein, *Remarks on the foundations of mathematics* (1983); Steven H. Holtzman and Christopher M. Leich, *Wittgenstein: To follow a rule* (1981); and Saul Kripke, *Wittgenstein on rules and private language* (1982).

24 Ludwig Wittgenstein, *Philosophical investigations* (1968: Part 2, 223).

2 ON SPACE IN GEOGRAPHIC INFORMATION SYSTEMS

1 See, for example, my discussion in Michael R. Curry, "On space and spatial practice in human geography" (1996), and the slightly different version in Michael R. Curry, *The work in the world: Geographical practice and the written word* (1996).

2 Lucretius, *On the nature of the universe* (1994).

3 Aristotle, *Physics* (1984).

4 Isaac Newton, *The mathematical principles of natural philosophy and his system of the world* (1934), Scholium §2.

5 H.G. Alexander, *The Leibniz-Clarke correspondence, together with extracts from Newton's Principia and Opticks* (1956); Gottfried Wilhelm Leibniz, *Monadology, and other philosophical essays* (1965).

6 Gottfried Wilhelm Leibniz, *Monadology, and other philosophical essays* (1965: fifth paper, §47).

7 Here see Samuel Y. Edgerton, *The Renaissance rediscovery of linear perspective* (1975); Martin Kemp, *The science of art: Optical themes in Western art from Brunelleschi to Seurat* (1990); John White, *The birth and rebirth of pictorial space* (1967).

8 Martin Kemp, *The science of art: Optical themes in Western art from Brunelleschi to Seurat* (1990: 12–13).

9 E.H. Gombrich, *Art and illusion: A study in the psychology of pictorial representation* (1956: 294).

10 Ptolemy, "The elements of geography" (1948: 162–3).

11 Svetlana Alpers, "The mapping impulse in Dutch art" (1987: 70).

12 Svetlana Alpers, "The mapping impulse in Dutch art" (1987: 71).

13 Ian L. McHarg, *Design with nature* (1969); Carl Steinitz, Paul Parker and Lawrie Jordan, "Hand-drawn overlays: Their history and prospective uses" (1976); and Warren Manning, "The Billerica town plan" (1913). My thanks to Francis Harvey for the references to Manning and Steinitz.

14 Christopher Alexander, Sara Ishikawa, and Murray Silverstein, *A pattern language: Towns, buildings, construction* (1977).

15 R.H. Atkin, *Mathematical structure in human affairs* (1974); R.H. Atkin, *Multidimensional man* (1981); Peter Gould, "Letting the data speak for themselves" (1981); Peter Gould, "Reflective distanciation through metamethodological perspective" (1983).

16 Ludwig Wittgenstein, *Tractatus logico-philosophicus* (1961: §2.18).

17 Ludwig Wittgenstein, *Tractatus logico-philosophicus* (1961: §2.2).

18 Ludwig Wittgenstein, *Tractatus logico-philosophicus* (1961: §2.202).

19 Norwood Russell Hanson, *Patterns of discovery* (1958).

20 John Krygier, "Envisioning the American west: Maps, the representational barrage of 19th century expedition reports, and the production of scientific knowledge" (1997).

21 Alexander von Humboldt, *Cosmos: A sketch of a physical description of the universe* (1813: vol. 2, 97–98). (As quoted in Krygier, above.)

3 OPTICAL CONSISTENCY, TECHNOLOGIES OF LOCATION, AND THE LIMITS OF REPRESENTATION

1 Bruno Latour, "Drawing things together" (1990).

2 David Gelernter, *Mirror worlds: Or the day software puts the universe in a shoebox: How it will happen and what it will mean* (1992: 3–17 passim).

3 Strabo, *The geography of Strabo* (1917: 3–5).

4 A useful popular summary, although now dated, is in John Noble Wilford, *The mapmakers: The story of the great pioneers in cartography from antiquity to the space age* (1981). The related issue of projections is covered in John Parr Snyder, *Flattening the Earth: Two thousand years of map projections* (1993).

5 The traditional process of establishing standards through triangulation is described in Mansfield Merriman, *An introduction to geodetic surveying. In three parts. I. The figure of the earth. II. The principles of least squares. III. The fieldwork of triangulation* (1892); F.R.

Gossett, *Manual of geodetic triangulation* (1950); US Coast and Geodetic Survey, *Geodesy: The transcontinental triangulation and the American arc of the parallel* (1900); and National Research Council Committee on the North American Datum, *North American datum: A report* (1971).

6 The basis for the newer system is described in National Research Council, Committee on Geodesy, *Geodesy: Trends and prospects* (1978) and Warren T. Dewhurst, *Input formats and specifications of the National Geodetic Survey data base* (1985).

7 See Charles S. Danner, Jr., "State Plane Coordinate System as a common horizontal datum for GIS and LIS systems" (1989).

8 Martin Baier, "Zip code – new tool for marketers" (1967: 140).

9 Martin Baier, "Zip code – new tool for marketers" (1967: 136).

10 Martin Baier, "Zip code – new tool for marketers" (1967: 136).

11 David J. Curry, *The new marketing research systems: How to use strategic database information for better marketing decisions* (1992); Erik Larson, *The naked consumer: How our private lives become public commodities* (1992); Michael J. Weiss, *The clustering of America* (1988) and *Latitudes and attitudes: An atlas of American tastes, trends, politics, and passions* (1994); Michael R. Curry, "Geodemographics and the end of the private realm" (1997); Jon Goss "We know who you are and we know where you live: The instrumental rationality of geo-marketing information systems" (1995).

12 "Fair credit reporting act of 1970" (1970)

13 Michael J. Weiss, *The clustering of America* (1988: 271).

14 Ludwig Wittgenstein, *Philosophical investigations* (1968: I, §217).

15 Ludwig Wittgenstein, *Philosophical investigations* (1968: II, 226).

16 Michael R. Curry, *The work in the world: Geographical practice and the written word* (1996); and Yi-Fu Tuan, *Space and place: The perspective of experience* (1977).

17 Eric Voegelin has showed that in the case of political bodies, this idea is a long-standing one; see Eric Voegelin, "The growth of the race idea" (1940).

18 David Lowenthal, *The past is a foreign country* (1985).

19 Vincent Berdoulay, "The Vidal-Durkheim debate" (1978).

20 Paul Vidal de la Blache, *The personality of France* (1928: 13). The introductory portion of his work, *Tableau de la géographie de la France* (1979) was translated into English and reprinted as *The personality of France*.

21 Paul Vidal de la Blache, *The personality of France* (1928: 14).

22 For additional views of the nature of geographical "personality," see Trevor J. Barnes and Michael R. Curry, "Towards a contextualist approach to geographical knowledge" (1983), and Gary S. Dunbar, "Geographical personality" (1974); in the latter it is argued that Vidal *et al.* have been much too stringent in their establishment of limitations on the application of the notion of "personality," and that judgments about the personality of places are constantly, and with no little justification, made.

23 Torsten Hagerstrand, "What about people in regional science?" (1970); Brian J.L. Berry, "Approaches to regional analysis: A synthesis" (1964).

24 Jan O. M. Broek, *The Santa Clara Valley, California: A study in landscape changes* (1932).

25 Michael F. Goodchild, "Stepping over the line: Technological constraints and the new cartography" (1988); Gail Langran, *Time in geographic information systems* (1992).

26 Alasdair C. MacIntyre, "Epistemological crises, dramatic narrative, and the philosophy of science" (1977); Louis O. Mink, "History and fiction as modes of comprehension" (1969) and "Narrative form as a cognitive instrument" (1978); Joseph Rouse, "The narrative reconstruction of science" (1990).

27 Polk Direct, "Niches from Polk Direct—promotional brochure," nd.

28 Robert Scholes, "Language, narrative and anti-narrative" (1981); Paul Ricoeur, "Narrative time" (1981).

29 In history, see Maurice Mandelbaum, "A note on history as narrative" (1967); William Dray, "On the nature and role of narrative in historiography" (1971); Hayden White, *Metahistory: The historical imagination in nineteenth-century Europe* (1973); Louis O. Mink, "Narrative form as a cognitive instrument" (1978); Allan Megill, "Recounting the past: Description, explanation, and narrative in historiography" (1989). In economics see Donald McCloskey, *The rhetoric of economics* (1985) and *If you're so smart: The narrative of economic expertise* (1990). In geography see Michael R. Curry and Trevor J. Barnes, "Time and narrative in economic geography" (1988); Andrew Sayer, "The new regional-geography and problems of narrative" (1989).

30 Arthur C. Danto, "Narrative sentences" (1962).

31 J. Nicholas Entrikin, *The betweenness of place* (1991); Thomas Nagel, *The view from nowhere* (1986).

32 Johannes Fabian, *Time and the other: How anthropology makes its object* (1983).

33 Yi-Fu Tuan, "Rootedness versus sense of place" (1980).

34 Bruce Kimball, *The 'true professional ideal' in America: A history* (1992).

35 Tom Conley, *The self-made map: Cartographic writing in early modern France* (1996); J.B. Harley, "Deconstructing the map" (1989) and "Maps, knowledge, and power" (1988).

36 Arthur O. Lovejoy, *The great chain of being: A study of the history of an idea* (1936); see also Paul G. Kuntz, "Hierarchy: From Lovejoy's great chain of being to Feibleman's great tree of being" (1971); Marion Leathers Kuntz and Paul G. Kuntz, *Jacob's ladder: Concepts of hierarchy and the great chain of being* (1988); Jaakko Hintikka and Heikki Kannisto, "Kant on 'the great chain of being' or the eventual realization of all possibilities: A comparative study" (1981).

37 Raymond Williams, *The country and the city* (1973).

4 ON THE ROOTS OF GEOGRAPHIC INFORMATION SYSTEMS

1 Preston James and Geoffrey Martin, *All possible worlds: A history of geographical ideas* (1981).

2 Arthur Robinson, Joel Morrison, Randall D. Sale, and Phillip C. Muehrcke, *Elements of cartography* (1995); and P.A. Burrough, *Principles of geographic information systems for land resource assessment* (1986). For a more general account of the phenomenon about which I am writing, see Henry Aay, "Textbook chronicles: Disciplinary history and the growth of geographic knowledge" (1981).

3 Langdon Winner, *Autonomous technology: Technics-out-of-control as a theme in political thought* (1977); and John Street, *Politics and technology* (1992).

4 However, see Donald Mackenzie, *Inventing accuracy: an historical sociology of nuclear missile guidance* (1990) for a critique of the notion that even in the case of ballistic missiles there is something independent, out there, that can be called "accuracy."

5 Actually, I would make a third, rather more complex, theoretical point, that the usual distinction between these "internal" histories and their "external" alternatives itself does not hold up under sustained critique. This point, which I have made elsewhere, suggests that the internal–external distinction rests upon an untenable theory of culture.

6 For a useful explication of this process, see Murray S. Davis, "'That's classic!' The phenomenology and rhetoric of successful social theories" (1986).

7 Two useful recent surveys are David Buisseret, *Monarchs, ministers, and maps: The emergence of cartography as a tool of government in early modern Europe* (1992) and Roger P. Kain and Elizabeth Baigent, *The cadastral map in the service of the state: A history of property mapping* (1992). In addition, I have found O.A.W. Dilke, *The Roman land surveyors: An introduction to the agrimensores* (1971) and *Greek and Roman maps* (1985), and Joseph Rykwert, *The idea of a town: The anthropology of urban form in Rome, Italy and the ancient world* (1988), to be useful introductions to the historical relationship between land surveying, mapping, and the state. An interesting case study is Denis Cosgrove, "The geometry of landscape: Practical and speculative arts in sixteenth-century Venetian land territories" (1988). The American experience is surveyed in two well-known works, Hildegard B. Johnson, *Order upon the land: The U.S. rectangular land survey and the upper Mississippi country* (1976) and Norman Thrower, *Original survey and land subdivision: A comparative study of the form and effect of contrasting cadastral systems* (1966).

8 Here, see Josef W. Konvitz, *Cartography in France: Science, engineering, and statecraft* (1987).

9 A set of articles in *The American Cartographer* lays out much of what I have termed the "standard" history. Especially interesting are David Rhind, "Personality as a factor in the development of a discipline: The example of computer-assisted cartography" (1988); J.T. Coppock, "The analogue to digital revolution: A view from an unreconstructed geographer" (1988); Jack Dangermond and Lowell Kent Smith, "Geographic information systems and the revolution in cartography: The nature of the role played by a commercial organization" (1988); Michael F. Goodchild, "Stepping over the line: Technological constraints and the new cartography" (1988); Nicholas R. Chrisman, "The risks of software innovation: A case study of the Harvard Lab" (1988); and Roger F. Tomlinson, "The impact of transition from analogue to digital cartographic representation" (1988).

10 David Rhind, "Personality as a factor in the development of a discipline: The example of computer-assisted cartography" (1988: 279).

11 Among the best known of the works of the early quantitative movement in geography are John Q. Stewart and William Warntz, "Macrogeography and social science" (1958); William Warntz, "Contributions toward a macroeconomic geography: A review" (1957); William Bunge, *Theoretical geography* (1966); and Fred, K. Schaefer, (1953).

12 Jack Dangermond and Lowell Kent Smith, "Geographic information systems and the revolution in cartography: The nature of the role played by a commercial organization" (1988).

5 THE RESHAPING OF GEOGRAPHIC PRACTICE

1 Robert K. Merton, "The normative structure of science" (1973: 275–76).

2 Robert K. Merton, "The normative structure of science" (1973: 273).

3 Barry S. Barnes and R.G.A. Dolby, "The scientific ethos: A deviant viewpoint" (1970); Walda Katz Fishman, "Science and society: Debunking the myth of scientific purity and autonomy" (1981).

4 Bruce Kimball, *The 'true professional ideal' in America: A history* 1992); see also Thomas L. Haskell, "Professionalism versus capitalism: R.H. Tawney, Emile Durkheim, and C.S. Peirce on the disinterestedness of professional communities" (1984).

5 Magali Sarfatti Larson, "Emblem and exception: The historical definition of the architect's professional role" (1983).

6 Indeed, Henry Petroski notes that when the pencil was first introduced it came with operating instructions; see Henry Petroski, *The pencil: A history of design and circumstance* (1990).

7 A provocative account of this process can be found in Harry M. Collins, "The replication of experiments in physics" (1982).

8 Terrence Toy and David Longbrake, "The begetting of a GIS laboratory: The University of Denver experience" (1994). Note that I do not mean to single out the University of Denver, but take their experience as being quite general.

9 David F. Noble, "Forces of production: A social history of industrial automation" (1984) and Shoshana Zuboff, *In the age of the smart machine: The future of work and power* (1988).

10 Here Barbara Garson gives ample evidence of a variety of cases in which the worker *is* treated as though he or she is an animal; see Barbara Garson, *The electronic sweatshop: How computers are transforming the office of the future into the factory of the past* (1988) and Barbara Garson, *All the livelong day: The meaning and demeaning of routine work* (1994).

11 Sherry Turkle and Seymour Papert, "Epistemological pluralism: Styles and voices within the computer culture" (1990).

12 Judy Wajcman, "Technology as masculine culture" (1991); and Langdon Winner, "Technologies as forms of life" (1986).

13 David Hounshell, *From the American system to mass production 1800–1932: The development of manufacturing technology in the United States* (1984); Thomas Parke Hughes, "The evolution of large technological systems" (1987); and Thomas Parke Hughes, *Networks of power: Electrification in western society, 1880–1930* (1983).

14 David F. Noble, *Forces of production: A social history of industrial automation* (1984). See also Barbara Garson, *All the livelong day: The meaning and demeaning of routine work* (1994); Barbara Garson, *The electronic sweatshop: How computers are transforming the office of the future into the factory of the past* (1988); and especially Shoshana Zuboff, *In the age of the smart machine: The future of work and power* (1988).

15 David Sudnow, *Pilgrim in the microworld* (1983).

16 Here Sudnow suggests that there are similarities between learning to play a video game—or write code—and learning to improvise in music; see his David Sudnow, *Ways of the hand: The organization of improvised conduct* (1978).

17 Michel Foucault, *Discipline and punish: The birth of the prison* (1977). Zuboff (1988) has provided an extremely interesting case study of this phenomenon in the case of the computerization of certain types of factory production.

18 Ivor Gratton-Guinness, "Work for the hairdressers: The production of de Prony's logarithmic and trigonometric tables" (1990). The account that follows, including the quotations, are from Martin Campbell-Kelly and William Aspray, *Computer: A history of the information machine* (1996). The title derives from the fact that most of those who did the computations were ex-hairdressers, who were forced to seek new work when under Napoleon hair styles changed.

19 Adam Smith, *An inquiry into the nature and causes of the wealth of nations* (1904:8).

20 Frederick W. Taylor, *Principles of scientific management* (1967); Frank B. Gilbreth, *Motion study: A method for increasing the efficiency of the workman* (1911).

21 Martin Campbell-Kelly and William Aspray, *Computer: A history of the information machine* (1996); for general histories of office technology, see JoAnne Yates, *Control through communication in American firms, 1850–1920* (1989) and James R. Beniger, *The control revolution: Technological and economic origins of the information society* (1986). And for a discussion of the relationship between human and computer reason, see Philip E. Agre, *Computation and human experience* (1997).

22 Harry Braverman, *Labor and monopoly capital: The degradation of work in the twentieth century* (1974).

23 Joan Greenbaum, "The head and the heart: Using gender analysis to study the social construction of computer systems" (1990: 9).

24 Evelyn Fox Keller, *Reflections on gender and science* (1985).

25 Michael R. Curry, "Forms of life and geographical method" (1989); Michael R. Curry, "The architectonic impulse and the reconceptualization of the concrete in contemporary geography" (1991); Langdon Winner, "Technologies as forms of life" (1986).

26 Alasdair C. MacIntyre, *After virtue: A study in moral theory* (1984).

27 Environmental Systems Research Institute, *Understanding GIS: The ARC/INFO Method (Version for UNIX and Windows NT)* (1996). Note that where a traditional book would be called the "sixth edition," the publishers here defer to the usage common to software, and refer to "Version 7.1." The work is described as follows:

> This revised workbook teaches the basics of GIS in the context of completing an ARC/INFO project. A series of hands-on exercises leads users through the steps of a typical project. The workbook is designed primarily for beginning ARC/INFO users, but GIS managers will find the first three lessons a useful introduction to the field. More experienced users can reference the book at various stages in their GIS projects. Understanding GIS can also provide the basis of the computer laboratory component for university GIS courses.

28 Tony Burns and Jim Henderson, "Education and training in GIS: The view from ESRI" (1989); and Anonymous, "University of Wisconsin-Madison offers ArcCAD courses" (1993).

29 See, to take a single example, David Green, David Rix, and James Cadoux-Hudson, *Geographic information 1994* (1994). Having singled it out, I should add that this particular volume contains a substantial number of useful essays.

30 See, for example, Thomas L. Haskell, "Professionalism versus capitalism: R.H. Tawney, Emile Durkheim, and C.S. Peirce on the disinterestedness of professional communities" (1984).

31 Elizabeth Eisenstein, *The printing press as an agent of change: Communications and cultural transformations in early modern Europe* (1979); Lucien Febvre and Henri Jean Martin, *The coming of the book: The impact of printing, 1450–1800* (1976); and Walter J. Ong, *Orality and literacy: The technologizing of the word* (1982).

32 Here see Tony Crowley, *Standard English and the politics of language* (1989); Aldo Scaglione, *The emergence of national languages* (1984); W. Haas, *Standard languages: Spoken and written* (1982); and James Milroy and Lesley Milroy, *Authority in language: investigating language prescription and standardisation* (1991). Some, as in Norman F. Blake, *Caxton and his world* (1969), have seen the case for a strict relationship between printing technology and standardization as far less compelling; here see also Thomas Cable, "The rise of written standard English" (1984).

33 Benedict Anderson, *Imagined communities: Reflections on the origin and spread of nation-
    alism* (1983). Here, see also Lucien Febvre and Henri Jean Martin, *The coming of the
    book: The impact of printing, 1450–1800* (1976) and Raymond Williams, *The long revolu-
    tion* (1961).

34 Arthur Robinson, "Mapmaking and map printing: The evolution of a working rela-
    tionship" (1975). The entire volume is to be recommended.

35 Arthur Robinson, "Mapmaking and map printing: The evolution of a working rela-
    tionship" (1975: 2).

36 The literature on the establishment of standards is, actually, surprisingly small. For a
    general account of the American experience, see Robert A. Martino, *Standardization
    activities of national technical and trade organizations* (1941). He notes that "The past
    quarter century has seen among other things the significant rise of standardization and
    simplification as methods of human progress" (p. 1), and goes on to note that by 1941
    there were in the United States 450 national technical societies and trade organiza-
    tions involved in the development of standards. Central, of course, was the National
    Bureau of Standards, which was not established until 1901, and the American Society
    for Testing Materials (ASTM), established in 1902. See also the popular John Perry,
    The story of standards (1955), and a useful set of essays in Dickson Reck, *National stan-
    dards in a modern economy* (1956).

37 Digital Cartographic Data Standards Task Force, "Proposed standard for digital carto-
    graphic data" (1988); Kathryn Neff, "The Spatial Data Transfer Standard (FIPS 173): A
    management perspective" (1992); and Robin G. Fegeas, Janette L. Cascio, and Robert
    A. Lazar, "An overview of FIPS 173, the Spatial Database Transfer Standard" (1992).

38 Digital Cartographic Data Standards Task Force, "Proposed standard for digital carto-
    graphic data" (1988).

39 The standard itself is National Institute of Standards and Technology, *Spatial Data
    Transfer Standard* (1992). It is discussed in varying detail in Nancy Tosta, "SDTS:
    Setting the standard" (1991); Henry Tom, "Spatial information and technology stan-
    dards evolving" (1992); Eric J. Strand, "A profile of GIS standards" (1992); Randy
    George, "The Spatial Data Transfer Standard" (1992); and H. J. Rossmeissl and R.
    D. Rugg, "An approach to data exchange: The Spatial Data Transfer Standard"
    (1992).

40 Much of what I say here draws from two documents, Maureen Breitenberg,
    "Questions and answers on quality, the ISO 9000 standard series, quality system
    registration, and related issues" (1991) and Maureen A. Breitenberg, "More questions
    and answers on the ISO 9000 standard series and related issues" (1993).

41 Office of Management and Budget "Circular A-16 (Revised): Coordination of
    surveying, mapping, and related spatial data activities" (1990).

42 National Institute of Standards and Technology (1992): *Spatial Data Transfer
    Standard*

6 WHO OWNS GEOGRAPHIC INFORMATION?

1 Strabo, *The geography of Strabo* (1917).

2 John Locke, *Second treatise of government* (1947: Ch. 5, Sec. 44).

3 G. W. F. Hegel, *Philosophy of right* (1967).

4 G. W. F. Hegel, *Philosophy of right* (1967: Paragraph 41).

5 G. W. F. Hegel, *Philosophy of right* (1967: Addition, Paragraph 41).

6 G. W. F. Hegel, *Philosophy of right* (1967: Paragraph 41).

7 Jane C. Ginsburg, "French copyright law: A comparative overview" (1989); Justin Hughes, "The philosophy of intellectual property" (1988); Arthur S. Katz, "The doctrine of moral rights and American copyright law—a proposal" (1951); Martin A. Roeder, "The doctrine of moral rights: A study in the law of artists, authors, and creators" (1940); Raymond Sarraute, "Current theory on the moral right of authors and artists under French law" (1968).

8 Karl Marx, *Capital* (1967); Frederick W. Taylor, *Principles of scientific management* (1967); Harry Braverman, *Labor and monopoly capital: The degradation of work in the twentieth century* (1974).

9 For an interesting account of this sociability, see Robert A. Leeson, *Travelling brothers: The six centuries' road from craft fellowship to trade unions* (1979).

10 John Locke, *Second treatise of government* (1947: Ch. 5, Sec. 27).

11 Sam Ricketson, *The Berne Convention for the protection of literary and artistic works: 1886–1986* (1987); Benjamin Kaplan, *An unhurried view of copyright* (1967); John Feather, "From rights in copies to copyright: The recognition of authors' rights in English law and practice in the sixteenth and seventeenth centuries" (1992); Justin Hughes, "The philosophy of intellectual property" (1988); Patrick Croskery, "The intellectual property literature: A structured approach" (1989); Vivian Weil and John W. Snapper, *Owning scientific and technical information: Value and ethical issues* (1989); Francis W. Rushing and Carole Ganz Brown, *Intellectual property rights in science, technology, and economic performance: International comparisons* (1990); J.B. Harley, "Maps, knowledge, and power" (1988) and "Deconstructing the map" (1989); and UNESCO, "Introductory report on 'scientists' rights" (1953).

12 Sam Ricketson, *The Berne Convention for the protection of literary and artistic works: 1886–1986* (1987: 5).

13 Martin A. Roeder, "The doctrine of moral rights: A study in the law of artists, authors, and creators" (1940); Pierre Masse, *Le droit moral de l'auteur sur son oeuvre littéraire ou artistique* (1906); Arthur S. Katz, "The doctrine of moral rights and American copyright law—a proposal" (1951); James M. Treece, "American law analogues of the author's 'moral right'" (1968); R.J. DaSilva, "'Droit moral' and the amoral copyright: A comparison of artists' rights in France and the U.S." (1980); and Rudolf Monta, "The concept of 'copyright' versus the 'droit d'auteur'" (1959).

14 Mark Rose, *Authors and owners: The invention of copyright* (1993).

15 Quoted in Jane C. Ginsburg, "A tale of two copyrights: Literary property in revolutionary France and America" (1990: 1873).

16 David N. Livingstone, *The geographical tradition* (1992).

17 Dava Sobel, *Longitude: The true story of a lone genius who solved the greatest scientific problem of his time* (1996); John Noble Wilford, "The matter of a degree" (1981).

18 Commission of the European Communities, "Council directive of 14 May 1991 on the legal protection of computer programs" (1991); and Commission of the European Communities, "Proposal for a council directive on the legal protection of databases" (1992).

19 The case was *Feist* v. *Rural Telephone*, 111 S. Ct. 1282 (1991). There have been a large number of works in response to Feist. Among them are: David Goldberg and Robert J. Bernstein, "The fallout from 'Feist': (Copyrightability of telephone listings)" (1991); Michael Schwartz, "Copyright in compilations of facts: Feist Publications, Inc. v. Rural Telephone Service" (1991); and Michael R. Klipper and Meredith S. Senter,

"The facts after Feist: The Supreme Court addresses the issue of the copyrightability of factual compilations" (1991). For the relationship between Feist and geographic information systems, see L.P. Dando, "Open records law, GIS, and copyright protection; Life after Feist" (1991).

20 I have drawn here on the analysis in VanGrasstek Communications, "Uruguay round: Further papers on selected issues" (1990).

21 General Agreement on Tariffs and Trade, "Final Act Embodying the Results of the Uruguay Round of Multilateral Trade Negotiations" (1993).

22 "Treaty with Poland concerning business and economic relations" (1990); and Letter from U. S. Trade Representative Carla A. Hills (1990).

23 Agreement on trade relations between the government of the United States of America and the government of Romania, and Side Letter to Honorable Constantin Fota, Minister of Commerce and Tourism, Romania, 3 April 1992.

24 Michael R. Curry, *The work in the world: Geographical practice and the written word* (1996: 163–74).

25 Edgar Zilsel, "The sociological roots of science" (1942).

26 Laura Kurgen, "You are here: Information drift" (1995: 42).

7 THE DIGITAL INDIVIDUAL IN A VISIBLE WORLD

1 Jeremy Crampton, "GIS and privacy: Crossing the line?" (1993); Jon Goss, "We know who you are and we know where you live: The instrumental rationality of Geo-Marketing Information Systems" (1993); Jon Goss, "Marketing the new marketing: The strategic discourse of Geodemographic Information Systems" (1994). Here I leave aside a variety of related issues. Of special interest are the issues of intelligent transportation systems and of electronic monitoring of released prisoners, both of which have strong geographic components. For the former see Philip E. Agre and Christine A. Harbs, "Social choice about privacy: Intelligent vehicle-highway systems in the United States" (1994) and Sheri Alpert, "Privacy and intelligent highways: Finding the right of way" (1994); for the latter see Ronald Corbett and Gary T. Marx, "Critique: No soul in the new machine: Technofallacies in the electronic monitoring movement" (1991) and Ann Aungles and David Cook, "Information technology and the family: Electronic surveillance and home imprisonment" (1994).

2 Lawrence M. Friedman, *The republic of choice: Law, authority, and culture* (1990). For more general accounts of privacy, see William Prosser, "Privacy [A legal analysis]" (1984); Edward J. Bloustein, "Privacy as an aspect of human dignity: An answer to Dean Prosser" (1964); J. Roland Pennock and John William Chapman, *Privacy* (1971); Stanley I. Benn and Gerald F. Gaus, "The public and the private: Concepts and action" (1983); Ferdinand Schoeman, "Privacy: Philosophical dimensions of the literature" (1984); Richard A. Posner, "An economic theory of privacy" (1984); Sissela Bok, *Secrets: On the ethics of concealment and revelation* (1983); Robert S. Gerstein, "Intimacy and privacy" (1984); Robert F. Murphy, "Social distance and the veil" (1984); Carole Pateman, "Feminist critiques of the public/private dichotomy" (1983); Jeffrey H. Reiman, "Privacy, intimacy, and personhood" (1984); and Alan F. Westin, *Privacy and freedom* (1967) and "The origins of modern claims to privacy" (1984).

3 Georg Simmel, "The metropolis and mental life" (1971); Louis Wirth, "Urbanism as a way of life" (1938); Louis Wirth, "Rural-urban differences" (1969).

4 Samuel Warren and Louis D. Brandeis, "The right of privacy" (1890).

5 David H. Flaherty, *Protecting privacy in surveillance societies: The Federal Republic of Germany, Sweden, France, Canada, and the United States* (1989: 8).

6 Herbert Fingarette, *On responsibility* (1967). The application of Fingarette's schema to the idea of privacy is my own.

7 Jeffrey H. Reiman, "Privacy, intimacy, and personhood" (1984); Ferdinand Schoeman, "Privacy: Philosophical dimensions of the literature" (1984).

8 These principles were laid out in U.S. Privacy Protection Study Commission, "Personal privacy in an information society" (1977).

9 Duncan Campbell and Steve Connor, *On the record: Surveillance, computers, and privacy: The inside story* (1986); Abbe Mowshowitz, *The conquest of will: Information processing in human affairs* (1976); Warren Freedman, *The right of privacy in the computer age* (1987); Jerry M. Rosenberg, *The death of privacy* (1969); David F. Linowes, *Privacy in America: Is your life in the public eye?* (1989); Joseph W. Eaton, *Card-carrying Americans: Privacy, security, and the national ID card debate* (1986); Kenneth C. Laudon, *Dossier society: Value choices in the design of national information systems* (1986).

10 Michael A. Epstein, Ronald S. Laurie, and Lawrence E. Elder, *Intellectual property: The European Community and Eastern Europe* (1993); David H. Flaherty, *Protecting privacy in surveillance societies: The Federal Republic of Germany, Sweden, France, Canada, and the United States* (1989); Colin J. Bennett, *Regulating privacy: Data protection and public policy in Europe and the United States* (1992).

11 Commission of the European Communities, "Commission communication on the protection of individuals in relation to the processing of personal data in the Community and information security" (1980); Council of Europe, "Protection of the privacy of individuals vis-à-vis electronic data banks in the private sector" (1973); Organisation for Economic Co-operation and Development, "Guidelines on the protection of privacy and transborder flows of personal data" (1981).

12 Council of Europe, "Protection of the privacy of individuals vis-à-vis electronic data banks in the private sector" (1973).

13 Council of Europe, "Convention for the protection of individuals with regard to automatic processing of data," 1981 #8447; see also Council of Europe, "Protection of personal data used for scientific research and statistics" (1984), "Protection of personal data used for the purposes of direct marketing" (1986), "Protection of personal data used for social security purposes" (1986), "Regulating the use of personal data in the police sector" (1988), "Protection of personal data used for employment purposes" (1989) and "New technologies: A challenge to privacy protection?" (1989).

14 Council of Europe, "Convention for the protection of individuals with regard to automatic processing of personal data" (1981: 21).

15 Organisation for Economic Co-operation and Development, "Guidelines on the protection of privacy and transborder flows of personal data" (1981: 9).

16 Organisation for Economic Co-operation and Development, "Guidelines on the protection of privacy and transborder flows of personal data" (1981: 11–12).

17 Commission of the European Communities, "Proposal for a council directive concerning the protection of individuals in relation to the processing of personal data" (1990). (Known as the "1990 Draft Directive".)

18 Commission of the European Communities, "Proposal for a council directive concerning the protection of individuals in relation to the processing of personal data" (1990: Ch. III, Arts. 8–12).

19 Commission of the European Communities, "Proposal for a council directive concerning the protection of individuals in relation to the processing of personal data" (1990, Ch. III, Art. 14).

20 Commission of the European Communities, "Proposal for a council directive concerning the protection of individuals in relation to the processing of personal data" (1990, Ch. III, Art. 15).

21 Commission of the European Communities, "Amended proposal for a council directive on the protection of individuals with regard to the processing of personal data and on the free movement of such data" (1992); see also Michael A. Epstein, Ronald S. Laurie, and Lawrence E. Elder, *Intellectual property: The European Community and Eastern Europe* (1993, IV, Sec. IV).

22 Council of the European Union, "Directive of the European Parliament and of the Council on the protection of individuals with regard to the processing of personal data and on the free movement of such data, as amended and approved 20 July 1995" (1995).

23 David H. Flaherty, *Privacy in colonial New England* (1967); see also Anthony G. Amsterdam, "Perspectives on the fourth amendment" (1974); Melvin Gutterman, "A formulation of the value and means models of the Fourth Amendment in the age of technologically enhanced surveillance" (1988); James J. Tomkovicz, "Beyond secrecy for secrecy's sake: Toward an expanded vision of the fourth amendment privacy province" (1985).

24 *Hester* v. *United States*. 265 US 57 (1924).

25 Anthony G. Amsterdam, "Perspectives on the fourth amendment" (1974).

26 In the analysis of court cases that follows, all of the decisions except *United States* v. *Penny-Feeney* were written by the United States Supreme Court.

27 *Boyd* v. *United States*. 116 US 616 (1886: 630).

28 *Olmstead* v. *United States*. 277 US 438 (1928).

29 *Olmstead* v. *United States*. 277 US 438 (1928: 4666).

30 *Hester* v. *United States*. 265 US 57 (1924).

31 *Katz* v. *United States*. 389 US 347 (1967).

32 *Smith* v. *Maryland*. 442 US 735 (1979).

33 *Katz* v. *United States*. 389 US 347 (1967: 361). So in fact we have here a two-part requirement. First, if police have gained evidence from a place that is an "open field," no search warrant is required, because of common law. And second, if the evidence has been gained from a place not an open field, from a home or the curtilage, for example, then the two-fold test laid out in *Katz* comes into play.

34 *Smith* v. *Maryland*. 442 US 735 (1979: 742).

35 *Smith* v. *Maryland*. 442 US 735 (1979: 743–44).

36 *United States* v. *Place*, 462 US 696 (1983).

37 Clifford S. Fishman, "Technologically enhanced visual surveillance and the Fourth Amendment: Sophistication, availability, and the expectation of privacy" (1988: 349–50).

38 But although it might be nice if court decisions were so firmly grounded in well-defined conceptual distinctions, a reading of the opinions and dissents in cases associated with search and seizure reveals something rather more unpleasant. One is

left with the distinct feeling that the majority decisions have been written with the intent of reaching an end, of getting drug dealers off the streets. One is left with the equally distinct impression that the arguments have been crafted solely with that end in mind, and that appeals to ideas have been used as means of persuasion. Nonetheless, it remains that the courts *have* been able with little dissent to appeal to a very general view of technological change.

39 Jacques Ellul, *The technological society* (1964); see also Langdon Winner, *Autonomous technology: Technics-out-of-control as a theme in political thought* (1977); John Street, *Politics and technology* (1992).

40 *Dow Chemical Co. v. United States*. 476 US 227 (1986); *California v. Ciraolo*. 476 US 207 (1986); *Florida v. Riley*. 488 US 445 (1989); *United States v. Penny-Feeney*. 773 F. Supp. 220 (D. Haw. 1991).

41 *California v. Ciraolo*. 476 US 207 (1986: 1812).

42 *California v. Ciraolo*. 476 US 207 (1986: 1813).

43 *Florida v. Riley*. 488 US 445 (1989: 451).

44 Here see also Lisa J. Steele, "The view from on high: Satellite remote sensing technology and the Fourth Amendment" (1991) and "Waste heat and garbage: The legalization of warrantless infrared searches" (1993).

45 *United States v. Knotts*. 460 US 276 (1983); *United States v. Karo*. 468 US 705 (1984).

46 *Florida v. Riley*. 488 US 445 (1989: 455).

47 Jennifer M. Granholm, "Video surveillance on public streets: The constitutionality of invisible citizen searches" (1987).

48 There is, no doubt, an underlying theory to geodemographics, and that theory surely derives in part from the Chicago School of urban sociology. Nonetheless, the development of geodemographic systems appears itself not to involve anything but the most cursory appeal to such theories.

49 "The Privacy Act of 1974" (1974); "Computer Matching and Privacy Protection Act of 1988" (1988); "Video Privacy Act of 1988" (1988).

50 Gary T. Marx and Nancy Reichman, "Routinizing the discovery of secrets" (1984).

51 Erik Larson, *The naked consumer: How our private lives become public commodities* (1992).

52 David H. Flaherty, *Protecting privacy in surveillance societies: The Federal Republic of Germany, Sweden, France, Canada, and the United States* (1989); Colin J. Bennett, *Regulating privacy: Data protection and public policy in Europe and the United States* (1992).

53 Lisa H. Greene and Steven J. Rizzi, "Database protection developments: Proposals stall in the United States and WIPO" (1997).

54 Columbia Human Rights Law Review, *Surveillance, dataveillance, and personal freedoms* (1973); James B. Rule, *Private lives and public surveillance* (1973); Alan F. Westin, *Databanks in a free society: Computers, record-keeping, and privacy* (1972).

55 Jeremy Bentham, *Panopticon; or, The inspection-house: containing the idea of a new principle of construction applicable to any sort of establishment, in which persons of any description are to be kept under inspection: and in particular to penitentiary-houses, prisons, houses of industry....   and schools: with a plan of management adapted to the principle* (1791). Note that the image of the Panopticon was popularized in Foucault's *Discipline and punish: The birth of the prison* (1977), but that Foucault's own argument, about the ways in which each person is a party to his or her own surveillance, both moves away from Bentham's image and seems particularly appropriate to the case of geodemographics.

56 G. Donald Bain, "Lotus primes MarketPlace for desktop marketing" (1991); Peter Huber, "Good tidings from Lotus development" (1990).

57 O'Connor, R.J. (1991) "Privacy flap kills Lotus data base," *San Jose Mercury News*, C1.

58 Mary J. Culnan, "The lessons of the Lotus MarketPlace: Implications for consumer privacy in the 1990s" (1991). Jim Seymour, "Lotus' MarketPlace succumbs to media hysteria" (1991); Laura J. Gurak, "Rhetorical dynamics of corporate communication in cyberspace: The protest over Lotus MarketPlace" (1995); Laura J. Gurak, "The rhetorical dynamics of a community protest in cyberspace: The case of Lotus MarketPlace." Ph.D. dissertation (1994).

59 David Gelernter, *Mirror worlds: Or the day software puts the universe in a shoebox: How it will happen and what it will mean* (1992).

60 Arthur Robinson, Joel Morrison, Randall D. Sale, and Phillip C. Muehrcke, *Elements of cartography* (1995); see also Borden Dent, "Visual organization and thematic map design" (1972); James J. Flannery, "The relative effectiveness of some common graduated point symbols in the presentation of quantitative data" (1971); Henry W. Castner and Arthur H. Robinson, *Dot area symbols in cartography: The influence of pattern on their perception* (1969).

61 *Mason v. Montgomery Data*. 967 F.2d 135 (5th Cir. 1992); David B. Wolf, "Is there copyright protection for maps after Feist?" (1992); David B. Wolf, "New landscape in the copyright protection for maps: Mason v. Montgomery Data, Inc." (1993). There have, of course, been more recent claims, that maps are not neutral or that they are ineluctably rhetorical; see J. B. Harley, "Silences and secrecy: The hidden agenda of cartography in early modern Europe" (1988); J.B. Harley, "Deconstructing the map" (1989), and Denis Wood, *The power of maps* (1992). But these claims, to the extent that they are directed against the assumption that people read maps directly, make it clear just how pervasive that realist view is.

62 Trans Union, "TIE—Trans Union's Income Estimator—Focusing on your true target—the individual" (1994).

63 Erving Goffman, *Forms of talk* (1981); Erving Goffman, *The presentation of self in everyday life* (1959).

64 Anthony Giddens, *Modernity and self-identity: Self and society in the late modern age* (1991); Pierre Bourdieu, "L'Illusion biographique" (1986); Sherry Turkle, *Life on the screen: Identity in the age of the Internet* (1995).

65 Berne Convention, "Berne Convention for the Protection of Literary and Artistic Works of September 9, 1886 . . . Amended on October 2, 1979" (1979). The United States, whose enacted law of intellectual property is based upon the Anglo-American theory of property, has been able successfully to argue that it incorporates the required moral-right provisions into its legal system because authors who suffer misrepresentation have recourse to the tort system.

66 Michael R. Curry, *The work in the world: Geographical practice and the written word* (1996), and "Data protection and intellectual property: Information systems and the Americanization of the new Europe" (1996); Jane C. Ginsburg, "French copyright law: A comparative overview" (1989); Arthur S. Katz, "The doctrine of moral rights and American copyright law—a proposal" (1951); Justin Hughes, "The philosophy of intellectual property" (1988); Martin A. Roeder, "The doctrine of moral rights: A study in the law of artists, authors, and creators" (1940); Raymond Sarraute, "Current theory on the moral right of authors and artists under French law" (1968).

8 GEOGRAPHIC INFORMATION SYSTEMS AND THE PROBLEM OF ETHICAL ACTION

1 Immanuel Kant, *Foundations of the metaphysics of morals* (1976).

2 Jeremy Bentham, *An introduction to the principles of morals and legislation* (1948); John Stuart Mill, *Utilitarianism* (1972).

3 David Hume, *Enquiries concerning human understanding and concerning the principles of morals* (1975).

4 Bertrand Russell, "Logical atomism" (1956); A.J. Ayer, *Language, truth, and logic* (1952); C.K. Ogden and I. A. Richards, *The meaning of meaning* (1923); Charles L. Stevenson, "The nature of ethical disagreement" (1948).

5 William Bunge, *Theoretical geography* (1973); David Harvey, *Explanation in geography* (1969).

6 René Descartes, "Meditations on first philosophy" (1971), "Rules for the direction of the mind" (1971), and "Discourse on the method" (1971).

7 Aristotle, "Ethica Nichomachea" (1941).

8 Philip E. Agre, "Beyond the mirror world: Privacy and the representational practices of computing" (1997).

9 Elizabeth Eisenstein, *The printing press as an agent of change: Communications and cultural transformations in early modern Europe* (1979); Jack Goody, *The domestication of the savage mind* (1977); Walter J. Ong, *Orality and literacy: The technologizing of the word* (1982); Michael T. Clanchy, *From memory to written record: England 1066–1307* (1993); Michael R. Curry, *The work in the world: Geographical practice and the written word* (1996).

10 Hesiod, *Works and days* (1988).

11 Jean Piaget, *The moral judgment of the child* (1965).

12 Michael Polanyi, *Personal knowledge: Towards a post-critical philosophy* (1958)

13 Thomas S. Kuhn, *The structure of scientific revolutions* (1970).

14 Donald Mackenzie, *Inventing accuracy: an historical sociology of nuclear missile guidance* (1990); see also Thomas S. Kuhn, "The function of measurement in modern physical science" (1961).

15 Ludwig Wittgenstein, *Philosophical investigations* (1968: I, §217).

16 Steven Shapin and Simon Schaffer, *Leviathan and the air pump: Hobbes, Boyle, and the experimental life* (1985); Bruno Latour and Steve Woolgar, *Laboratory life: The social construction of scientific facts* (1979); Sharon Traweek, *Beamtimes and lifetimes: The world of high energy physicists* (1988).

17 Alasdair C. MacIntyre, *After virtue: A study in moral theory* (1984).

18 Richard Sclove, "Panel on the case against computers: A systemic critique" (1995); see also his *Democracy and technology* (1995).

9 BEYOND PaleoGIS?

1 See *Cartography and Geographic Information Systems,* Vol. 22, 1 (1995), and John Pickles, *Ground truth: The social implications of geographic information systems* (1995).

2 Neil Smith, "History and philosophy of geography: Real wars, theory wars" (1992); Robert W. Lake, "Planning and applied geography: Positivism, ethics, and geographic information systems" (1993).

3 As, for example, in Thomas Stearns Eliot, *Notes towards the definition of culture* (1949).

4 Aldo Leopold, *A Sand County almanac, with other essays on conservation* (1966).
5 Donald Janelle, "Spatial reorganization: A model and a concept" (1969); David Harvey, *The condition of postmodernity: An enquiry into the origins of cultural change* (1989).

BIBLIOGRAPHY

Aarsleff, H., *From Locke to Saussure: Essays on the study of language and intellectual history*, Minneapolis, MN: University of Minnesota Press, 1982.

Aay, H., "Textbook chronicles: Disciplinary history and the growth of geographic knowledge," in *The origins of academic geography in the United States*, ed. B.W. Blouet and T.L. Stitcher, 291–302, Hamden, CT: Archon Books, 1981.

Agre, P. and Harbs, C.A., "Social choice about privacy: Intelligent vehicle-highway systems in the United States," *Information Technology and People* 7, 4 (1994): 63–90.

Agre, P.E., "Beyond the mirror world: Privacy and the representational practices of computing," in *Technology and privacy: The new landscape*, ed. P.E. Agre and M. Rotenberg, 29–62, Cambridge, MA: MIT Press, 1997.

—— *Computation and human experience*, Cambridge: Cambridge University Press, 1997.

"Agreement on trade relations between the government of the United States of America and the government of Romania, and side letter to Honorable Constantin Fota, Minister of Commerce and Tourism, Romania, April 3, 1992." Copy available from the US Trade Representative, Executive Office of the President, Washington DC 20506, 1992.

Alexander, C., Ishikawa, S. and Silverstein, M., *A pattern language: Towns, buildings, construction*, ed. M. Jacobson, I. Fiksdahl-King, and S. Angel, New York: Oxford University Press, 1977.

Alexander, H.G., *The Leibniz-Clarke correspondence, together with extracts from Newton's Principia and Opticks*, New York: Barnes & Noble Imports, 1956.

Alpers, S., "The mapping impulse in Dutch art," in *Art and cartography: Six historical essays*, ed. D. Woodward, 51–96, Chicago, IL: University of Chicago Press, 1987.

Alpert, S., "Privacy and intelligent highways: Finding the right of way," *Santa Clara Computer and High Technology Law Journal* 11, 1 (1994): 97–118.

Amsterdam, A.G., "Perspectives on the fourth amendment," *Minnesota Law Review* 58 (1974): 349–477.

Anderson, B., *Imagined communities: Reflections on the origin and spread of nationalism*, London: Verso, 1983.

Anonymous, "University of Wisconsin-Madison offers ArcCAD courses," *ARC News*, Winter (1993): 11.

Aristotle, "Ethica Nichomachea," in *The basic works of Aristotle*, ed. R. McKeon, 935–1126, New York: Random House, 1941.

——, *Physics*, trans. J. Barnes. *The complete works*, Vol. I, 315–46. Princeton, NJ: Princeton University Press, 1984.

Arnauld, A. and Lancelot, C., *The Port-Royal grammar: General and rational grammar*, trans. J. Rieux and B.E. Rollin, The Hague: Mouton, 1975.

Atkin, R.H., *Mathematical structure in human affairs*, New York: Crane, Rusak, & Co., 1974.

—— *Multidimensional man*, Harmondsworth: Penguin, 1981.

Aungles, A. and Cook, D., "Information technology and the family: Electronic surveillance and home imprisonment," *Information Technology and People* 7, 1 (1994): 69–80.

Ayer, A.J., *Language, truth, and logic*, 2nd edn., New York: Dover, 1952.

Baier, M., "Zip code – new tool for marketers," *Harvard Business Review* 45, 1 (1967): 136–40.

Bain, G.D., "Lotus primes MarketPlace for desktop marketing," *MacWEEK* 5, 3 (1991): 31ff.

Barnes, B.S. and Dolby, R.G.A., "The scientific ethos: A deviant viewpoint," *Archives Europeans de sociologie* 11, 1 (1970): 3–25.

Barnes, T.J. and Curry, M.R., "Towards a contextualist approach to geographical knowledge," *Transactions, Institute of British Geographers* NS 8 (1983): 467–82.

Beniger, J.R., *The control revolution: Technological and economic origins of the information society*, Cambridge, MA: Harvard University Press, 1986.

Benn, S.I. and Gaus, G.F., "The public and the private: Concepts and action," in *Public and private in social life*, 3–27, London: Croom Helm, 1983.

Bennett, C.J., *Regulating privacy: Data protection and public policy in Europe and the United States*, Ithaca, NY: Cornell University Press, 1992.

Bentham, J., *Panopticon; or, The inspection-house: containing the idea of a new principle of construction applicable to any sort of establishment, in which persons of any description are to be kept under inspection: and in particular to penitentiary-houses, prisons, houses of industry. . . and schools: with a plan of management adapted to the principle*, 3 vols, London: T. Payne, 1791.

—— *An introduction to the principles of morals and legislation*, New York: Hafner, 1948.

Berdoulay, V., "The Vidal-Durkheim debate," in *Humanistic geography: Prospects and problems*, ed. D. Ley and M. Samuels, 77–90, Chicago, IL: Maaroufa Press, 1978.

Berne Convention for the Protection of Literary and Artistic Works: message from the President of the United States transmitting the Berne Convention for the Protection of Literary and Artistic Works of September 9, 1886, completed at Paris on May 4, 1896, revised at Berlin on November 13, 1908, completed at Berne on March 20, 1914, and revised at Rome on June 2, 1928, at Brussels on June 26, 1948, at Stockholm on July 14, 1967, and at Paris on July 24, 1971 and amended in 1979. Vol. 99–127., Treaty doc. Washington: US. G.P.O. 1986.

Berry, B.J.L., "Approaches to regional analysis: A synthesis," *Annals of the Association of American Geographers* 54 (1964): 2–11.

Blades, M., "Navigation and wayfinding in information systems," in *Human factors in geographical information systems*, ed. D. Medyckyj-Scott and H.M. Hearnshaw, 61–9, London: Bellhaven Press, 1993.

Blake, N.F., *Caxton and his world*, New York: House & Maxwell, 1969.

Bloor, D., *Wittgenstein: A social theory of knowledge*, New York: Columbia University Press, 1983.

Bloustein, E.J., "Privacy as an aspect of human dignity: An answer to Dean Prosser," *New York University Law Review* 39 (1964): 962–1007.

Bok, S., *Secrets: On the ethics of concealment and revelation*, New York: Random House, 1983.

Bourdieu, P., "L'Illusion biographique," *Actes de la recherche en sciences sociales* 62–3, June (1986): 69–72.

Boyd v. *United States, 116 US 616 (1886)*, 1886.

Braverman, H., *Labor and monopoly capital: The degradation of work in the twentieth century*, New York: Monthly Review Press, 1974.

Breitenberg, M. A., "Questions and answers on quality, the ISO 9000 standard series, quality system registration, and related issues." Washington, DC: US Department of Commerce, 1991.

—— "More questions and answers on the ISO 9000 standard series and related issues." Washington, DC: US Department of Commerce, 1993.

Broek, J.O.M., *The Santa Clara Valley, California: A study in landscape changes*, Utrecht: Oosthoek, 1932.

Buisseret, D., *Monarchs, ministers, and maps: The emergence of cartography as a tool of government in early modern Europe*, Chicago, IL: University of Chicago Press, 1992.

Bunge, W., *Theoretical geography*, Lund: C.W.K. Gleerup, 1966.

—— *Theoretical geography*, 2nd revised and enlarged edn, Lund: C.W.K. Gleerup, 1973.

Burns, T. and Henderson, J., "Education and training in GIS: The view from ESRI," Paper presented at Auto-Carto 9: Ninth international symposium on computer-assisted cartography, Falls Church, VA, April 2–7, 1989.

Burrough, P.A., *Principles of geographic information systems for land resource assessment*, New York: Oxford University Press, 1986.

Cable, T., "The rise of written standard English," in *The emergence of national languages*, ed. A. Scaglione, 75–94, Ravenna: Longo, 1984.

California v. *Ciraolo, 476 US 207 (1986)*, 1986.

Campbell, D. and Connor, S., *On the record: Surveillance, computers, and privacy: The inside story*, London: M. Joseph, 1986.

Campbell-Kelly, M. and Aspray, W., *Computer: A history of the information machine*, New York: Basic Books, 1996.

Carnap, R., *The logical structure of the world: Pseudoproblems in philosophy*, trans. R.A. George, Berkeley, CA: University of California Press, 1967.

Castner, H.W. and Robinson, A.H., *Dot area symbols in cartography: The influence of pattern on their perception*, Washington, DC: American Congress of Surveying & Mapping, 1969.

Chomsky, N., *Rules and representations*, New York: Columbia University Press, 1980.

Chrisman, N.R., "The risks of software innovation: A case study of the Harvard Lab," *The American Cartographer* 15 (1988): 291–300.

Clanchy, M.T., *From memory to written record: England 1066–1307*, 2nd edn, Oxford: Blackwell, 1993.

Collins, H.M., "The replication of experiments in physics," in *Science in context: Readings in the sociology of science*, ed. B. Barnes and D. Edge, 94–116, Cambridge, MA: MIT Press, 1982.

Columbia Human Rights Law Review, *Surveillance, dataveillance, and personal freedoms*, Fair Lawn, NJ: R.E. Burdick, 1973.

Commission of the European Communities, "Commission communication on the protection of individuals in relation to the processing of personal data in the Community and information security," Brussels: European Community, 1980.

—— "Proposal for a council directive concerning the protection of individuals in relation to the processing of personal data," Brussels: European Community, 1990.

—— "Council directive of 14 May 1991 on the legal protection of computer programs," Brussels: European Community, 1991.

—— "Amended proposal for a council directive on the protection of individuals with regard to the processing of personal data and on the free movement of such data," Brussels: European Community, 1992.

—— "Proposal for a council directive on the legal protection of databases," Brussels: European Community, 1992.

Computer Matching and Privacy Protection Act of 1988. US 1988. H.R. 4699, 100th Cong., 2d sess.

Conley, T., *The self-made map: Cartographic writing in early modern France*, Minneapolis, MN: University of Minnesota Press, 1996.

Coppock, J.T., "The analogue to digital revolution: A view from an unreconstructed geographer," *The American Cartographer* 15, 3 (1988): 263–75.

Corbett, R. and Marx, G.T., "Critique: No soul in the new machine: Technofallacies in the electronic monitoring movement," *Justice Quarterly* 8, 3 (1991): 400–14.

Cosgrove, D., "The geometry of landscape: Practical and speculative arts in sixteenth-century Venetian land territories," in *The iconography of landscape: Essays on the symbolic representation, design, and use of past environments*, ed. D. Cosgrove and S. Daniels, 254–76, Cambridge: Cambridge University Press, 1988.

Council of Europe, "Protection of the privacy of individuals vis-à-vis electronic data banks in the private sector," Strasbourg: Council of Europe, 1973.

—— "Convention for the protection of individuals with regard to automatic processing of personal data," Strasbourg: Council of Europe, 1981.

—— "Protection of personal data used for scientific research and statistics," Strasbourg: Council of Europe, 1984.

—— "Protection of personal data used for social security purposes," Strasbourg: Council of Europe, 1986.

—— "Protection of personal data used for the purposes of direct marketing," Strasbourg: Council of Europe, 1986.

—— "Regulating the use of personal data in the police sector," Strasbourg: Council of Europe, 1988.

—— "New technologies: A challenge to privacy protection?," Strasbourg: Council of Europe, 1989.

—— "Protection of personal data used for employment purposes," Strasbourg: Council of Europe, 1989.

Council of the European Union, "Directive of the European Parliament and of the Council on the protection of individuals with regard to the processing of personal data and on the free movement of such data, as amended and approved 20 July 1995," Brussels: European Parliament and Council of the European Union, 1995.

Crampton, J., "GIS and privacy: Crossing the line?," paper presented at the workshop on geographic information and society, Friday Harbor, WA, 1993.

Croskery, P., "The intellectual property literature: A structured approach," in *Owning scientific and technical information: Value and ethical issues*, ed. V. Weil and J.W. Snapper, 268–82, New Brunswick, NJ: Rutgers University Press, 1989.

Crowley, T., *Standard English and the politics of language*, Urbana, IL: University of Illinois Press, 1989.

Culnan, M.J., "The lessons of the Lotus MarketPlace: Implications for consumer privacy in the 1990s," Paper presented at the First Conference on Computers, Freedom, and Privacy, Burlingame, CA, 1991.

Curry, D.J., *The new marketing research systems: How to use strategic database information for better marketing decisions*, New York: John Wiley and Sons, 1992.

Curry, M.R., "Forms of life and geographical method," *Geographical Review* 79, 3 (1989): 280–96.

—— "The architectonic impulse and the reconceptualization of the concrete in contemporary geography," in *Writing worlds: Discourse, text, and metaphor in the representation of landscape*, ed. J. Duncan and T.J. Barnes, 97–117, New York: Routledge, 1991.

—— "Data protection and intellectual property: Information systems and the Americanization of the new Europe," *Environment and Planning A* 28 (1996): 891–908.

—— "On space and spatial practice in human geography," in *Concepts in human geography*, ed. C. Earle, K. Mathewson, and M. Kenzer, 3–32, New York: Rowman & Allenheld, 1996.

—— *The work in the world: Geographical practice and the written word*, Minneapolis, MN: University of Minnesota Press, 1996.

—— "The digital individual and the private realm," *Annals, Association of American Geographers* 87, 4 (1997): 681–99.

Curry, M.R. and Barnes, T.J., "Time and narrative in economic geography," *Environment and Planning A* 20 (1988): 141–9.

Dando, L.P., "Open records law, GIS, and copyright protection: Life after Feist," *URISA Proceedings* (1991): 1–17.

Dangermond, J. and Smith, L.K., "Geographic information systems and the revolution in cartography: The nature of the role played by a commercial organization," *The American Cartographer* 15, 3 (1988): 301–10.

Danner, C.S., Jr., "State Plane Coordinate System as a common horizontal datum for GIS and LIS systems," Paper presented at the ASPRS-ACSM Fall Convention, Falls Church, VA 1989.

Danto, A.C., "Narrative sentences," *History and Theory* 2 (1962): 146–79.

DaSilva, R.J., "'Droit moral' and the amoral copyright: A comparison of artists' rights in France and the U. S," *Bulletin of the Copyright Society of the USA* 28, 1 (1980): 1–58.

Davis, M.S., "'That's classic!' The phenomenology and rhetoric of successful social theories," *Philosophy of the Social Sciences* 16 (1986): 285–301.

Davis, P.J. and Hersh, R., "Rhetoric and mathematics," in *The rhetoric of the human sciences: Language and argument in scholarship and public affairs*, ed. J.S. Nelson, A. Megill, and D. McCloskey, 53–68, Madison, WI: University of Wisconsin Press, 1987.

Dent, B., "Visual organization and thematic map design," *Annals of the Association of American Geographers* 62 (1972): 79–93.

Descartes, R., "Discourse on the method," in *Descartes' philosophical writings*, 5–58, Indianapolis: Bobbs-Merrill, 1971.

—— "Meditations on first philosophy," in *Descartes' philosophical writings*, 59–124, Indianapolis: Bobbs-Merrill, 1971.

—— "Rules for the direction of the mind," in *Descartes' philosophical writings*, 151–80, Indianapolis: Bobbs-Merrill, 1971.

Dewhurst, W.T., *Input formats and specifications of the National Geodetic Survey data base*, Rockville, MD: National Oceanic and Atmospheric Administration, 1985.

Digital Cartographic Data Standards Task Force, "Proposed standard for digital cartographic data," *The American Cartographer* 15, 1 (1988): 7–140.

Dilke, O.A.W., *The Roman land surveyors: An introduction to the agrimensores*, New York: Barnes & Noble, 1971.

—— *Greek and Roman maps*, Ithaca, NY: Cornell University Press, 1985.

Dow Chemical Co. v. United States., 476 US 227 (1986), 1986.

Dray, W., "On the nature and role of narrative in historiography," *History and Theory* 10 (1971): 153–71.

Dunbar, G.S., "Geographical personality," *Geoscience and Man* 5 (1974): 25–33.

Eaton, J.W., *Card-carrying Americans: Privacy, security, and the national ID card debate*, Totowa, NJ: Rowman & Littlefield, 1986.

Edgerton, S.Y., *The Renaissance rediscovery of linear perspective*, New York: Harper & Row, 1975.

Eisenstein, E., *The printing press as an agent of change: Communications and cultural transformations in early modern Europe*, Cambridge: Cambridge University Press, 1979.

Eliot, T.S., *Notes towards the definition of culture*, New York: Harcourt, Brace, & Co., 1949.

Ellul, J., *The technological society*, trans. J. Wilkinson, New York: Random House, 1964.

Entrikin, J.N., *The betweenness of place*, Basingstoke: Macmillan, 1991.

Environmental Systems Research Institute, *Understanding GIS: The ARC/INFO Method (Version for UNIX and Windows NT)*, Version 7.1, addendum included Redlands, CA: ESRI, 1996.

Epstein, M.A., Laurie, R.S., and Elder, L.E., *Intellectual property: The European Community and Eastern Europe*, Englewood Cliffs, NJ: Prentice Hall, 1993.

Fabian, J., *Time and the other: How anthropology makes its object*, New York: Columbia University Press, 1983.

Fair Credit Reporting Act of 1970. US 1970. 15 USC Sec. 1681.

Feather, J., "From rights in copies to copyright: The recognition of authors' rights in English law and practice in the sixteenth and seventeenth centuries," *Cardozo Arts and Entertainment Law Journal* 10, 2 (1992): 455–74.

Febvre, L. and Martin, H.J., *The coming of the book: The impact of printing, 1450–1800*, trans. D. Gerard, ed. G. Nowell-Smith and D. Wooton, London: NLB, 1976.

Fegeas, R.G., Cascio, J.L. and Lazar, R.A., "An overview of FIPS 173, the Spatial Database Transfer Standard," *Cartography and Geographic Information Systems* 19, 5 (1992).

Feist v. *Rural Telephone, 111 S. Ct. 1282 (1991)*, 1991.

Fingarette, H., *On responsibility*, New York: Basic Books, 1967.

Fishman, C.S., "Technologically enhanced visual surveillance and the Fourth Amendment: Sophistication, availability, and the expectation of privacy," *American Criminal Law Review* 26 (1988): 315–58.

Fishman, W.K., "Science and society: Debunking the myth of scientific purity and autonomy," *Humanity and Society* 5, 2 (1981): 140–65.

Flaherty, D.H., *Privacy in colonial New England*, Charlottesville, VA: University Press of Virginia, 1967.

—— *Protecting privacy in surveillance societies: The Federal Republic of Germany, Sweden, France, Canada, and the United States*, Chapel Hill, NC: The University of North Carolina Press, 1989.

Flannery, J.J., "The relative effectiveness of some common graduated point symbols in the presentation of quantitative data," *The Canadian Cartographer* 8, 2 (1971): 96–109.

Florida v. *Riley, 488 U S 445 (1989)*, 1989.

Foucault, M., *Discipline and punish: The birth of the prison*, trans. A. Sheridan, New York: Vintage Books, 1977.

Freedman, W., *The right of privacy in the computer age*, New York: Quorum Books, 1987.

Frege, G., "On sense and reference," in *Translations from the philosophical writings of Gottlob Frege*, 56–78. Oxford: Basil Blackwell, 1952.

—— *Translations from the philosophical writings of Gottlob Frege*, trans. M. Black, Oxford: Basil Blackwell, 1952.

Friedman, L.M., *The republic of choice: Law, authority, and culture*, Cambridge, MA: Harvard University Press, 1990.

Garnham, A., *Artificial intelligence: An introduction*, London: Routledge, 1987.

Garson, B., *The electronic sweatshop: How computers are transforming the office of the future into the factory of the past*, New York: Simon & Schuster, 1988.

—— *All the livelong day: The meaning and demeaning of routine work*, revised edn, New York: Penguin, 1994.

Gavison, R., "Information control: Availability and exclusion," in *Public and private in social life*, ed. S.I. Benn and G.F. Gaus, 113–34, London: Croom Helm, 1983.

Gelernter, D., *Mirror worlds: Or the day software puts the universe in a shoebox: How it will happen and what it will mean*, New York: Oxford University Press, 1992.

General agreement on Tariffs and Trade, "Final Act Embodying the Results of the Uruguay Round of Multilateral Trade Negotiations." Genera: GATT Secretariat, 1993.

George, R., "The Spatial Data Transfer Standard," in *1991–92 international GIS sourcebook*, ed. H. Dennison Parker, 425–26, Fort Collins, CO: GIS World, 1992.

Gerstein, R.S., "Intimacy and privacy," in *Philosophical dimensions of privacy: An anthology*, ed. F. Schoeman, 265–71, Cambridge: Cambridge University Press, 1984.

Giddens, A., *Modernity and self-identity: Self and society in the late modern age*, Stanford, CA: Stanford University Press, 1991.

Gilbreth, F.B., *Motion study: A method for increasing the efficiency of the workman*, New York: Van Nostrand, 1911.

Ginsburg, J.C., "French copyright law: A comparative overview," *Journal of the Copyright Society of the USA* 4 (1989): 269–85.

—— "A tale of two copyrights: Literary property in revolutionary France and America," *Tulane Law Review* 64 (1990): 991–1031.

Gluck, M., "Making sense of human wayfinding: Review of cognitive and linguistic knowledge for personal navigation with a new research direction," in *Cognitive and linguistic aspects of geographic space: An introduction*, ed. D.M. Mark and A.U. Frank, 117–36, Dordrecht: Kluwer Academic, 1991.

Goffman, E., *The presentation of self in everyday life*, Garden City, NY: Doubleday Anchor, 1959.

—— *Forms of talk*, Philadelphia, PA: University of Pennsylvania Press, 1981.

Goldberg, D. and Bernstein, R.J., "The fallout from 'Feist': (Copyrightability of telephone listings)," *New York Law Journal* 206, 63 (1991): 3ff.

Gombrich, E.H., *Art and illusion: A study in the psychology of pictorial representation*, Princeton, NJ: Princeton University Press, 1956.

Goodchild, M.F., "Stepping over the line: Technological constraints and the new cartography," *The American Cartographer* 15, 3 (1988): 311–19.

Goody, J., *The domestication of the savage mind*, Cambridge: Cambridge University Press, 1977.

Goss, J., "We know who you are and we know where you live: The instrumental rationality of Geo-Marketing Information Systems," Paper presented at the Workshop on geographic information and society, Friday Harbor, WA 1993.

—— "Marketing the new marketing: The strategic discourse of geodemographic information systems," in *Ground truth: The social implications of geographic information systems*, ed. J. Pickles, 130–70, New York: Guilford Press, 1994.

—— "We know who you are and we know where you live: The instrumental rationality of Geo-Marketing Information Systems," *Economic Geography* 71, 2 (1995): 171–98.

Gossett, F.R., *Manual of geodetic triangulation*, Washington, DC: US Government Printing Office, 1950.

Gould, P., "Letting the data speak for themselves," *Annals of the Association of American Geographers* 71 (1981): 166–76.

—— "Reflective distanciation through metamethodological perspective," *Environment and Planning B* 10 (1983): 381–92.

Granholm, J.M., "Video surveillance on public streets: The constitutionality of invisible citizen searches," *University of Detroit Law Review* 64 (1987): 687–713.

175

Gratton-Guinness, I., "Work for the hairdressers: The production of de Prony's logarithmic and trigonometric tables," *Annals of the History of Computing* 12, 3 (1990): 177–85.

Green, D., Rix, D. and Cadoux-Hudson, J., *Geographic information 1994*, London: Taylor & Francis, in conjunction with the Association for Geographic Information, 1994.

Greenbaum, J., "The head and the heart: Using gender analysis to study the social construction of computer systems," *Computers and Society* 20, 2 (1990): 9–18.

Greene, L.H. and Rizzi, S.J., "Database protection developments: Proposals stall in the United States and WIPO," *Journal of proprietary rights* 9, 1 (1997): 2–8.

Gurak, L.J., "The rhetorical dynamics of a community protest in cyberspace: The case of Lotus MarketPlace," Ph.D. dissertation, Rensselaer Polytechnic Institute, 1994.

—— "Rhetorical dynamics of corporate communication in cyberspace: The protest over Lotus MarketPlace," *IEEE Transactions on Professional Communication* 38, 1 (1995): 2–10.

Gutterman, M., "A formulation of the value and means models of the Fourth Amendment in the age of technologically enhanced surveillance," *Syracuse Law Review* 39 (1988): 647–736.

Haas, W., *Standard languages: Spoken and written*, Manchester: Manchester University Press, 1982.

Hagerstrand, T., "What about people in regional science?" *Papers of the Regional Science Association* 24 (1970): 7–21.

Hall, G.B., Wang, F., and Subaryono, "Comparison of Boolean and fuzzy classification methods in land suitability analysis by using geographical information systems," *Environment and Planning A* 24, 4 (1992): 497–516.

Hanson, N.R., *Patterns of discovery*, Cambridge: Cambridge University Press, 1958.

Harley, J.B., "Maps, knowledge, and power," in *The iconography of landscape: Essays on the symbolic representation, design, and use of past environments*, ed. D. Cosgrove and S. Daniels, 277–312, Cambridge: Cambridge University Press, 1988.

—— "Silences and secrecy: The hidden agenda of cartography in early modern Europe," *Imago Mundi* 40 (1988): 57–76.

—— "Deconstructing the map," *Cartographica* 26, 2 (1989): 1–20.

Harvey, D., *Explanation in geography*, London: Edward Arnold, 1969.

—— *The condition of postmodernity: An enquiry into the origins of cultural change*, New York: Basil Blackwell, 1989.

Haskell, T.L., "Professionalism versus capitalism: R.H. Tawney, Emile Durkheim, and C.S. Peirce on the disinterestedness of professional communities," in *The authority of experts: Studies in history and theory*, ed. T.L. Haskell, 180–225, Bloomington, IN: Indiana University Press, 1984.

Hegel, G.W.F., *Philosophy of right*, trans. T.M. Knox, London: Oxford University Press, 1967.

Hesiod, "Works and days," in *Theogony and Works and days* 35–61. Oxford: Oxford University Press, 1988.

Hester v. United States, 265 U S 57 (1924), 1924.

Hintikka, J. and Kannisto, H., "Kant on 'the great chain of being' or the eventual realiza-tion of all possibilities: A comparative study," in *Reforging the great chain of being: studies of the history of modal theories*, ed. S. Knuuttila, 287–308, Dordrecht: D. Reidel, 1981.

Holtzman, S.H. and Leich, C.M., *Wittgenstein: To follow a rule*, London: Routledge & Kegan Paul, 1981.

Hounshell, D., *From the American system to mass production 1800–1932: The development of manufacturing technology in the United States*, Baltimore, MD: Johns Hopkins University Press, 1984.

Huber, P., "Good tidings from Lotus development," *Forbes* 146, 14 (1990): 136.

Hughes, J., "The philosophy of intellectual property," *Georgetown Law Journal* 77, 2 (1988): 287–366.

Hughes, T.P., *Networks of power: Electrification in western society, 1880–1930*, Baltimore, MD: Johns Hopkins University Press, 1983.

—— "The evolution of large technological systems," in *The social construction of technolog-ical systems: New directions in the sociology and history of technology*, ed. W.E. Bijker, T.P. Hughes, and T.J. Pinch, 51–82, Cambridge, MA: MIT Press, 1987.

Humboldt, A. von., *Cosmos: A sketch of a physical description of the universe*, trans. E.C. Ott, London: Henry G. Bohn, 1813.

Hume, D., *Enquiries concerning human understanding and concerning the principles of morals*, 3rd edn, Oxford: Oxford University Press, 1975.

James, P. and Martin, G., *All possible worlds: A history of geographical ideas*, 2nd edn, New York: Wiley, 1981.

Janelle, D., "Spatial reorganization: A model and a concept," *Annals of the Association of American Geographers* 59 (1969): 348–64.

Johnson, H.B., *Order upon the land: The U. S. rectangular land survey and the upper Mississippi country*, New York: Oxford University Press, 1976.

Kain, R.P. and Baigent, E., *The cadastral map in the service of the state: A history of property mapping*, Chicago, IL: University of Chicago Press, 1992.

Kant, I., *Foundations of the metaphysics of morals*, trans. L.W. Beck, Indianapolis, IN: Bobbs Merrill, 1976.

Kaplan, B., *An unhurried view of copyright*, New York: Columbia University Press, 1967.

Katz, A.S., "The doctrine of moral rights and American copyright law—a proposal," *Southern California Law Review* 24 (1951): 375–427.

Katz v. United States, 389 U S 347 (1967), 1967.

Keller, E.F., *Reflections on gender and science*, New Haven, CT: Yale University Press, 1985.

Kemp, M., *The science of art: Optical themes in Western art from Brunelleschi to Seurat*, New Haven, CT: Yale University Press, 1990.

Kim, T.J., Wiggins, L.L. and Wright, J.R., *Expert systems: Applications in urban planning*, New York: Springer-Verlag, 1990.

Kimball, B., *The 'true professional ideal' in America: A history*, Cambridge: Blackwells, 1992.

Klipper, M.R. and Senter, M.S., "The facts after Feist: The Supreme Court addresses the issue of the copyrightability of factual compilations," in *Fact and data protection after Feist*, ed. J.A. Baumgarten, Englewood Cliffs, NJ: Prentice Hall, 1991.

Konvitz, J.W., *Cartography in France: Science, engineering, and statecraft*, Chicago, IL: University of Chicago Press, 1987.

Kripke, S., *Wittgenstein on rules and private language*, Cambridge, MA: Harvard University Press, 1982.

Krygier, J., "Envisioning the American West: Maps, the representational barrage of 19th century expedition reports, and the production of scientific knowledge," *Cartography and Geographic Information Systems* 24, 1 (1997): 27–50.

Kuhn, T.S., "The function of measurement in modern physical science," *Isis* 52 (1961): 162–76.

—— *The structure of scientific revolutions*, 2nd, enlarged edn, Chicago, IL: University of Chicago Press, 1970.

Kuntz, M.L. and Kuntz, P.G., *Jacob's ladder: Concepts of hierarchy and the great chain of being*, revised edn, New York: Peter Lang, 1988.

Kuntz, P.G., "Hierarchy: From Lovejoy's great chain of being to Feibleman's great tree of being" *Studium Generale* 24 (1971): 678–87.

Kurgen, L., "You are here: Information drift," *Assemblage* 25 (1995): 15–43.

Lake, R.W., "Planning and applied geography: Positivism, ethics, and geographic information systems," *Progress in Human Geography* 17, 3 (1993): 404–13.

Lakoff, G., *Women, fire, and dangerous things: What categories reveal about the mind*, Chicago, IL: University of Chicago Press, 1987.

Lakoff, G. and Johnson, M., *Metaphors we live by*, Chicago, IL: University of Chicago Press, 1980.

Langran, G., *Time in geographic information systems*, Bristol, PA: Taylor & Francis, 1992.

Larson, E., *The naked consumer: How our private lives become public commodities*, New York: Penguin, 1992.

Larson, M.S., "Emblem and exception: The historical definition of the architect's professional role," in *Professionals and urban form*, ed. J.R. Blau, M.E. La Gory, and J.S. Pipkin, 49–86, Albany, NY: State University of New York Press, 1983.

Latour, B., "Drawing things together," in *Representation in scientific practice*, ed. M. Lynch and S. Woolgar, 19–68, Cambridge, MA: MIT Press, 1990.

Latour, B. and Woolgar, S., *Laboratory life: The social construction of scientific facts*, Beverly Hills: Sage Publications, 1979.

Laudon, K.C., *Dossier society: Value choices in the design of national information systems*, New York: Columbia University Press, 1986.

Le Corbusier, *Towards a new architecture*, trans. F. Etchells, New York: Dover, 1986.

Leeson, R.A., *Travelling brothers: The six centuries' road from craft fellowship to trade unions*, London: Allen & Unwin, 1979.

Leibniz, G.W., *Monadology, and other philosophical essays*, trans. P. Schrecker and A.M. Schrecker, Indianapolis: Bobbs-Merrill, 1965.

Leopold, A., *A Sand County almanac, with other essays on conservation*, New York: Oxford University Press, 1966.

Letter from US Trade Representative Carla A. Hills, 1990.

Leung, L., "A prospectus based on fuzzy logic and knowledge-based geographical information systems," *Asian Geographer* 9, 1 (1990): 1–9.

Levi-Strauss, C., *The savage mind*, Chicago, IL: University of Chicago Press, 1968.

Linowes, D.F., *Privacy in America: Is your life in the public eye?* Urbana, IL: University of Illinois Press, 1989.

Livingstone, D.N., *The geographical tradition*, Oxford: Blackewll, 1992.

Lobeck, A.K., *Block diagrams and other graphic methods used in geology and geography*, 2nd edn, Amherst, MA: Emerson-Trussell Book Co., 1958.

Locke, J., Second treatise of government, in *Two treatises of government*, 121–247, New York: Hafner, 1947.

—— *An essay concerning human understanding*, Oxford: Oxford University Press, 1975.

Lovejoy, A.O., *The great chain of being: A study of the history of an idea*, Cambridge, MA: Harvard University Press, 1936.

Lowenthal, D., *The past is a foreign country*, New York: Cambridge University Press, 1985.

Lucretius, *On the nature of the universe*, trans. R.E. Latham, revised J. Godwin, revised edn, Harmondsworth: Penguin, 1994.

McCloskey, D., *The rhetoric of economics*, Madison, WI: University of Wisconsin Press, 1985.

—— *If you're so smart: The narrative of economic expertise*, Chicago, IL: University of Chicago Press, 1990.

McHarg, I.L., *Design with nature*, Garden City, NY: Published for the American Museum of Natural History by the Natural History Press, 1969.

MacIntyre, A.C., "Epistemological crises, dramatic narrative, and the philosophy of science," *Monist* 60 (1977): 453–72.

—— *After virtue: A study in moral theory*, 2nd edn, Notre Dame, IL: University of Notre Dame Press, 1984.

Mackenzie, D., *Inventing accuracy: an historical sociology of nuclear missile guidance*, Cambridge, MA: MIT Press, 1990.

Malcolm, N., *Ludwig Wittgenstein: A memoir*, London: Oxford University Press, 1966.

Mandelbaum, M., "A note on history as narrative," *History and Theory* 6 (1967): 413–19.

Manning, W., "The Billerica town plan," *Landscape Architecture* 3, 3 (1913): 108–18.

Mark, D.M., "Representation of geographic space in natural language, minds, culture, and computers," *Proceedings, Conference of Latin Americanist Geographers* (1990).

Martino, R.A., *Standardization activities of national technical and trade organizations*, Washington, DC: US Government Printing Office, 1941.

Marx, G.T. and Reichman, N., "Routinizing the discovery of secrets," *American Behavioral Scientist* 27, 4 (1984): 423–52.

Marx, K., *Capital*, New York: International Publishers, 1967.

Mason v. Montgomery Data, 967 F.2d 135 (5th Cir. 1992), 1992.

Masse, P., *Le droit moral de l'auteur sur son oeuvre littéraire ou artistique*, Paris: Arthur Rousseau, 1906.

Megill, A., "Recounting the past: Description, explanation, and narrative in historiography," *American Historical Review* 94, 3 (1989): 627–53.

Merchant, A., "Expert system: A design methodology," *Computers, Environment, and Urban Systems* 16, 1 (1992): 21–41.

Merriman, M., *An introduction to geodetic surveying. In three parts: I. The figure of the earth. II. The principles of least squares. III. The fieldwork of triangulation*, New York: T. Wiley & Sons, 1892.

Merton, R.K., "The normative structure of science," in *The sociology of science: Theoretical and empirical investigations*, ed. N.W. Storer, 267–78, Chicago, IL: University of Chicago Press, 1973.

Mill, J.S., *Utilitarianism*, London: Dent, 1972.

Milroy, J. and Milroy, L., *Authority in language: investigating language prescription and standardisation*, London: Routledge, 1991.

Mink, L.O., "History and fiction as modes of comprehension," *New Literary History* 1 (1969): 541–58.

—— "Narrative form as a cognitive instrument," in *The writing of history: Literary form and historical understanding*, ed. R.H. Canary and H. Kozicki, 129–49, Madison, WI: The University of Wisconsin Press, 1978.

Monta, R., "The concept of 'copyright' versus the 'droit d'auteur'," *Southern California Law Review* 32 (1959): 177–86.

Mowshowitz, A., *The conquest of will: Information processing in human affairs*, Reading, MA: Addison-Wesley, 1976.

Murphy, R.F., "Social distance and the veil," in *Philosophical dimensions of privacy: An anthology*, ed. F. Schoeman, 34–55, Cambridge: Cambridge University Press, 1984.

Nagel, T., *The view from nowhere*, New York: Oxford University Press, 1986.

National Institute of Standards and Technology. *Spatial Data Transfer Standard*, Gaithersburg, MD: Computer Systems Laboratory, National Institute of Standards and Technology, 1992.

National Research Council, Committee on the North American Datum, *North American datum: A report*, Washington, DC: National Academy of Science, 1971.

National Research Council, Committee on Geodesy, *Geodesy: Trends and prospects*, Washington, DC: National Academy of Science, 1978.

Neff, K., "The Spatial Data Transfer Standard (FIPS 173): A management perspective," *Cartography and Geographic Information Systems* 19, 5 (1992).

Newell, A., Shaw, J.C. and Simon, H.A., "Empirical explorations with the logic theory machine: A case study in heuristics," *Proceedings of the Western Joint Computer Conference* 15 (1957): 218–30.

Newton, I., *The mathematical principles of natural philosophy and his system of the world*, trans. F. Cajori, Berkeley, CA: University of California Press, 1934.

Noble, D.F., *Forces of production: A social history of industrial automation*, New York: Oxford University Press, 1984.

O'Connor, R.J., "Privacy flap kills Lotus data base," *San Jose Mercury News*, Jan. 24, 1991, C1.

Office of Mangement and Budget Circular A-16 (Revised): Coordination of surveying, mapping, and related spatial data activities 1990.

Ogden, C.K. and Richards, I.A., *The meaning of meaning*, London: Kegan Paul, 1923.

Olmstead v. United States, 277 U S 438 (1928), 1928.

O'Malley, M., *Keeping watch: A history of American time*, New York: Penguin, 1990.

Ong, W. J., *Orality and literacy: The technologizing of the word*, London: Routledge, 1982.

Organisation for Economic Co-operation and Development, "Guidelines on the protection of privacy and transborder flows of personal data," Paris: OECD, 1981.

Pateman, C., "Feminist critiques of the public/private dichotomy," in *Public and private in social life*, ed. S.I. Benn and G.F. Gaus, 281–303, London: Croom Helm, 1983.

Pearl, J., *Heuristics: Intelligent strategies for computer problem solving*, Reading, MA: Addison-Wesley, 1984.

Pennock, J.R. and Chapman, J.W., *Privacy*, New York: Atherton Press, 1971.

Perry, J., *The story of standards*, New York: Funk & Wagnells, 1955.

Petroski, H., *The pencil: A history of design and circumstance*, New York: Knopf, 1990.

Piaget, J., *The moral judgment of the child*, trans. M. Gabin, New York: The Free Press, 1965.

Pickles, J., *Ground truth: The social implications of geographic information systems*, New York: Guilford Press, 1995.

"Plans unveiled for major release of ARC/INFO," *ARC News*, Summer 1996, 6.

Plato., *Theaetetus*, trans. F.M. Cornford, Indianapolis, IN: Bobbs-Merrill.

—— "The sophist," in *Plato's theory of knowledge*, ed. Cornford, F.M., 165–332, London: Routledge & Kegan Paul, 1935.

—— *The republic*, trans. F.M. Cornford, New York: Oxford University Press, 1945.

Polanyi, M., *Personal knowledge: Towards a post-critical philosophy*, London: Routledge & Kegan Paul, 1958.

—— *The tacit dimension*, Gloucester, MA: Peter Smith, 1983.

Polk Direct, "Niches from Polk Direct—promotional brochure," nd.

Posner, R.A., "An economic theory of privacy," in *Philosophical dimensions of privacy: An anthology*, ed. F. Schoeman, 333–45, Cambridge: Cambridge University Press, 1984.

Pred, A., "Production, family, and free-time productions: A time-geographic perspective on the individual and societal change in nineteenth-century U. S. cities," *Journal of Historical Geography* 7, 1 (1981): 3–36.

The Privacy Act of 1974. U. S. 1974. PL 93–579, 5 USC 552a, Sec. 3(e) (4).

Prosser, W., "Privacy [A legal analysis]," in *Philosophical dimensions of privacy: An anthology*, ed. F. Schoeman, 104–55, Cambridge: Cambridge University Press, 1984.

Ptolemy, "The elements of geography," in *A source book in Greek science*, ed. M.R. Cohen, 162–81, Cambridge, MA: Harvard University Press, 1948.

Rachels, J., "Why privacy is important," in *Ethical issues in the use of computers*, ed. D.G. Johnson and J.W. Snapper, 194–201, Belmont, CA: Wadsworth Publishing, 1985.

Reck, D., *National standards in a modern economy*, New York: Harper, 1956.

Reiman, J.H., "Privacy, intimacy, and personhood," in *Philosophical dimensions of privacy: An anthology*, ed. F. Schoeman, 300–16, Cambridge: Cambridge University Press, 1984.

Reiss, T.J., *The discourse of modernism*, Ithaca, NY: Cornell University Press, 1982.

Rhind, D., "Personality as a factor in the development of a discipline: The example of computer-assisted cartography," *The American Cartographer* 15, 3 (1988): 277–89.

Ricketson, S., *The Berne Convention for the protection of literary and artistic works: 1886–1986*, London: Centre for Commercial Law Studies, Queen Mary College, 1987.

Ricoeur, P., "Narrative time," in *On narrative*, ed. W.J.T. Mitchell, 165–86, Chicago, IL: University of Chicago Press, 1981.

Robinson, A., "Mapmaking and map printing: The evolution of a working relationship," in *Five centuries of map printing*, ed. D. Woodward, 1–24, Chicago, IL: University of Chicago Press, 1975.

Robinson, A., Morrison, J., Sale, R.D. and Muehrcke, P.C., *Elements of cartography*, 6th edn, New York: John Wiley & Sons, 1995.

Roeder, M.A., "The doctrine of moral rights: A study in the law of artists, authors, and creators," *Harvard Law Review* 53 (1940): 554–78.

Rose, M., *Authors and owners: The invention of copyright*, Cambridge: Harvard University Press, 1993.

Rosenberg, J.M., *The death of privacy*, New York: Random House, 1969.

Rossmeissl, H.J. and Rugg, R.D., "An approach to data exchange: The Spatial Data Transfer Standard," in *Geographic information systems (GIS) and mapping—practices and standards*, ed. A.I. Johnson, C.B. Pettersson, and J.L. Fulton, 38–44, Philadelphia, PA: American Society for Testing and Materials, 1992.

Rouse, J., "The narrative reconstruction of science," *Inquiry* 33 (1990): 179–90.

Rule, J.B., *Private lives and public surveillance*, London: Allen Lane, 1973.

Rushing, F.W. and Brown, C.G., *Intellectual property rights in science, technology, and economic performance: International comparisons*, Boulder: Westview Press, 1990.

Russell, B., "Descriptions", in *Introduction to mathematical philosophy*, 167–80, London: Allen & Unwin, 1920.

—— "Logical atomism," in *Logic and knowledge: Essays 1901–1950*, ed. R.C. Marsh, 175–82, London: George Allen & Unwin, 1956.

—— "On denoting," in *Logic and knowledge: Essays 1901–1950*, ed. R.C. Marsh, 41–56, London: George Allen & Unwin, 1956.

Rykwert, J., *The idea of a town: The anthropology of urban form in Rome, Italy and the ancient world*, Cambridge: MIT Press, 1988.

Ryle, G., *The concept of mind*, New York: Barnes & Noble, 1949.

St. Augustine, *Confessions*, trans. R.S. Pine-Coffin, New York: Penguin, 1961.

Sarraute, R., "Current theory on the moral right of authors and artists under French law," *The American Journal of Comparative Law* 16 (1968): 465–86.

Sayer, A., "The new regional-geography and problems of narrative," *Environment and Planning D: Society and Space* 7, 3 (1989): 253–76.

Scaglione, A., *The emergence of national languages*, Ravenna: Longo, 1984.

Schaefer, F.K., "Exceptionalism in geography: A methodological examination," *Annals of the Association of American Geographers*, 43 (1953): 226–49.

Schoeman, F., "Privacy: Philosophical dimensions of the literature," in *Philosophical dimensions of privacy: An anthology*, 1–33, Cambridge: Cambridge University Press, 1984.

Scholes, R., "Language, narrative and anti-narrative," in *On narrative*, ed. W.J.T. Mitchell, 200–8, Chicago, IL: University of Chicago Press, 1981.

Schulman, R.D., "Portable GIS: From the sands of Desert Storm to the forests of California," *Geo Info Systems* 1, 8 (1991): 24.

Schwartz, M., "Copyright in compilations of facts: Feist Publications, Inc. v. Rural Telephone Service," *European Intellectual Property Review* (1991): 178–82.

Sclove, R., *Democracy and technology*, New York: Guilford Press, 1995.

—— "Panel on the case against computers: A systemic critique," Paper presented at the Fifth Conference on Computers, Freedom, and Privacy, Burlingame, CA 1995.

Seymour, J., "Lotus' MarketPlace succumbs to media hysteria," *PC Week* 8, 5 (1991): 57.

Shapin, S. and Schaffer, S., *Leviathan and the air pump: Hobbes, Boyle, and the experimental life*, Princeton, NJ: Princeton University Press, 1985.

Simmel, G., "The metropolis and mental life," in *Georg Simmel on individuality and social forms*, 324–39, Chicago, IL: University of Chicago Press, 1971.

Slaughter, M., *Universal languages and scientific taxonomy in the seventeenth century*, Cambridge: Cambridge University Press, 1982.

Smith, A., *An inquiry into the nature and causes of the wealth of nations*, London: Methuen, 1904.

Smith, N., "History and philosophy of geography: Real wars, theory wars," *Progress in Human Geography* 16, 2 (1992): 257–71.

Smith v. *Maryland, 442 US 735 (1979)*, 1979.

Snyder, J.P., *Flattening the Earth: Two thousand years of map projections*, Chicago, IL: University of Chicago Press, 1993.

Sobel, D., *Longitude: The true story of a lone genius who solved the greatest scientific problem of his time*: Penguin, 1996.

Soja, E.W., *Postmodern geographies: The reassertion of space in critical social theory*, London: Verso, 1989.

Steele, L.J., "The view from on high: Satellite remote sensing technology and the Fourth Amendment," *High Technology Law Journal* 6, 2 (1991): 317–34.

—— "Waste heat and garbage: The legalization of warrantless infrared searches," *Criminal Law Bulletin* 29 (1993): 19–39.

Steinitz, C., Parker, P. and Jordan, L., "Hand-drawn overlays: Their history and prospective uses," *Landscape Architecture* (1976): 444–55.

Stevenson, C.L., "The nature of ethical disagreement," *Sigma* 2 (1948): 469–76.

Stewart, J.Q. and Warntz, W., "Macrogeography and social science," *Geographical Review* 48 (1958): 167–84.

Strabo, *The geography of Strabo*, trans. H.L. Jones, London: Heinemann, 1917.

Strand, E.J., "A profile of GIS standards," in *1991–92 international GIS sourcebook*, ed. H. Dennison Parker, 417–21, Fort Collins, CO: GIS World, 1992.

Street, J., *Politics and technology*, Basingstoke: Macmillan, 1992.

Sudnow, D., *Ways of the hand: The organization of improvised conduct*, Cambridge: Harvard University Press, 1978.

—— *Pilgrim in the microworld*, New York: Warner Books, 1983.

Sui, D.Z., "A fuzzy GIS modeling approach for urban land evaluation," *Computers, Environment, and Urban Systems* 16, 2 (1992): 101–15.

Taylor, F.W., *Principles of scientific management*, New York: Norton, 1967.

Thrower, N., *Original survey and land subdivision: A comparative study of the form and effect of contrasting cadastral systems*, Chicago, IL: Rand McNally, 1966.

Tom, H., "Spatial information and technology standards evolving," in *1991–92 international GIS sourcebook*, ed. H. Dennison Parker, 422–6, Fort Collins, CO: GIS World, 1992.

Tomkovicz, J.J., "Beyond secrecy for secrecy's sake: Toward an expanded vision of the fourth amendment privacy province," *The Hastings Law Journal* 36, 5 (1985): 645–738.

Tomlinson, R.F., "The impact of transition from analogue to digital cartographic representation," *The American Cartographer* 15 (1988): 249–61.

Tosta, N., "SDTS: Setting the standard," *Geo Info Systems* 1, 7 (1991): 57–9.

Toy, T. and Longbrake, D., "The begetting of a GIS laboratory: The University of Denver experience," *ARC News*, Spring (1994): 19.

Trans Union, "TIE—Trans Union's Income Estimator—Focusing on your true target—the individual," 1994.

Traweek, S., *Beamtimes and lifetimes: The world of high energy physicists*, Cambridge, MA: Harvard University Press, 1988.

"Treaty with Poland concerning business and economic relations: message from the President of the United States transmitting the treaty between the United States of America and the Republic of Poland concerning business and economic relations with protocol and four related exchanges of letters signed March 21, 1990, at Washington." Washington, DC: U.S. Government Printing Office, 1990.

Treece, J.M., "American law analogues of the author's 'moral right'," *The American Journal of Comparative Law* 16 (1968): 487–506.

Tuan, Yi-Fu, *Space and place: The perspective of experience*, Minneapolis, MN: University of Minnesota Press, 1977.

—— "Rootedness versus sense of place," *Landscape* 24, 1 (1980): 3–8.

Turing, A.M., "Computing machinery and intelligence," *Mind* 59 (1950): 433–60.

Turkle, S., *Life on the screen: Identity in the age of the Internet*, New York: Simon & Schuster, 1995.

Turkle, S. and Papert, S., "Epistemological pluralism: Styles and voices within the computer culture," *Signs: Journal of Women in Culture and Society* 16, 1 (1990): 128–57.

Treaty with Poland concerning business and economic relations. US 1990. US Treaty Doc. 18, 101st Congress, 2d Session (1990).

US Coast and Geodetic Survey, *Geodesy: The transcontinental triangulation and the American arc of the parallel*, Washington, DC: US Government Printing Office, 1900.

Privacy Protection Study Commission, "Personal privacy in an information society," Washington, DC: US Government Printing Office, 1977.

UNESCO, "Introductory report on 'scientists' rights," 1954.

United States v. Karo, 468 US 705 (1984), 1984.

United States v. Knotts, 460 US 276 (1983), 1983.

United States v. Penny-Feeney, 773 F. Supp. 220 (D. Haw. 1991), 1991.

United States v. Place, 462 US 696, 1983.

VanGrasstek Communications, "Uruguay round: Further papers on selected issues," New York: United Nations, 1990.

Vidal de la Blache, P., *The personality of France*, trans. H.C. Brentnall, London: Alfred A. Knopf, 1928.

—— *Tableau de la géographie de la France*, Paris: Tallandier, 1979.

Video Privacy Act of 1988. US 1988. 18 USC Sec. 2901.

Voegelin, E., "The growth of the race idea," *The Review of Politics* 2 (1940): 283–317.

Wajcman, J., "Technology as masculine culture," in *Feminism confronts technology*, 137–61, University Park, PA: Pennsylvania State University Press, 1991.

Warntz, W., "Contributions toward a macroeconomic geography: A review," *Geographical Review* 47 (1957): 420–4.

Warren, S. and Brandeis, L.D., "The right of privacy," *Harvard Law Review* 4, 5 (1890): 193–220.

Wasserstrom, R.A., "Privacy: Some arguments and assumptions," in *Philosophical dimensions of privacy: An anthology*, ed. F. Schoeman, 317–32, Cambridge: Cambridge University Press, 1984.

Waterman, D.A., *A guide to expert systems*, Reading, MA: Addison-Wesley, 1986.

Weil, V. and Snapper, J.W., *Owning scientific and technical information: Value and ethical issues*, New Brunswick, NJ: Rutgers University Press, 1989.

Weiss, M.J., *The clustering of America*, New York: Harper & Row, 1988.

—— *Latitudes and attitudes: An atlas of American tastes, trends, politics, and passions*, Boston, MA: Little, Brown, & Co., 1994.

Westin, A.F., *Privacy and freedom*, New York: Atheneum, 1967.

—— *Databanks in a free society: Computers, record-keeping, and privacy*, New York: Quadrangle Books, 1972.

—— "The origins of modern claims to privacy," in *Philosophical dimensions of privacy: An anthology*, ed. F. Schoeman, 56–74, Cambridge: Cambridge University Press, 1984.

White, H., *Metahistory: The historical imagination in nineteenth-century Europe*, Baltimore, MD: Johns Hopkins University Press, 1973.

White, J., *The birth and rebirth of pictorial space*, 2nd edn, London: Faber & Faber, 1967.

White, M., "Car navigation systems," in *Geographical Information Systems: principles and applications*, ed. D.J. Maguire, M.F. Goodchild and D.W. Rhind, 115–25, Essex: Longman Scientific & Technical, 1991.

Whitehead, A.N. and Russell, B., *Principia mathematica*, 3 vols, Cambridge: The University Press, 1925–7.

Wilford, J.N., *The mapmakers: The story of the great pioneers in cartography from antiquity to the space age*, New York: Random House, 1981.

—— "The matter of a degree," in *The mapmakers: The story of the great pioneers in cartography from antiquity to the space age*, 92–110, New York: Random House, 1981.

Wilkins, J., *An essay towards a real character, and a philosophical language*, Menston, Yorkshire: Scolar Press, 1668.

Williams, R., *The long revolution*, New York: Columbia University Press, 1961.

—— *The country and the city*, New York: Oxford University Press, 1973.

Winner, L., *Autonomous technology: Technics-out-of-control as a theme in political thought*, Cambridge, MA: MIT Press, 1977.

—— "Technologies as forms of life," in *The whale and the reactor: A search for limits in an age of high technology*, 3–18, Chicago, IL: University of Chicago Press, 1986.

Wirth, L., "Urbanism as a way of life," *American Journal of Sociology* 44 (1938): 1–24.

—— "Rural-urban differences," in *Classic essays in the culture of cities*, ed. R. Sennett, 165–9, New York: Appleton-Century-Crofts, 1969.

Wittgenstein, L., *Tractatus logico-philosophicus*, trans. D.F. Pears and B.F. McGuinness, London: Routledge & Kegan Paul, 1961.

—— *Philosophical investigations*, trans. G.E.M Anscombe, 3rd edn, New York: Macmillan, 1968.

—— *Remarks on the foundations of mathematics*, trans. G. H. Von Wright, R. Rhees, and G.E.M. Anscombe, revised edn, Cambridge, MA: MIT Press, 1983.

Wolf, D.B., "Is there copyright protection for maps after Feist?," *Journal of the Copyright Society of the USA* 39, 3 (1992): 224–42.

—— "New landscape in the copyright protection for maps: Mason v. Montgomery Data, Inc.," *Journal of the Copyright Society of the USA* 40, 3 (1993): 401–6.

Wood, D., *The power of maps*, New York: Guilford, 1992.

Yates, J., *Control through communication in American firms, 1850–1920*, Baltimore, MD: Johns Hopkins University Press, 1989.

Zilsel, E., "The sociological roots of science," *American Journal of Sociology* 47 (1942): 544–62.

Zuboff, S., *In the age of the smart machine: The future of work and power*, New York: Basic Books, 1988.

INDEX